Settlement Ecology

Arizona Studies in Human Ecology

Glenn Davis Stone

Settlement Ecology

The Social and Spatial Organization of Kofyar Agriculture

The University of Arizona Press / Tucson

The University of Arizona Press

∞ This book is printed on acid-free, archival-quality paper
Manufactured in the United States of America
First printing

Library of Congress Cataloging-in-Publication Data
Stone, Glenn Davis.
Settlement ecology: the social and spatial organization of Kofyar agriculture/
Glenn Davis Stone.
p. cm.—(Arizona studies in human ecology)
Includes bibliographical references and index.
ISBN 0-8165-1567-0 (cloth: acid-free paper)
1. Kofyar (African people)—Agriculture. 2. Kofyar (African people)—Land tenure. 3. Kofyar (African people)—Social conditions. 4. Land settlement patterns—Nigeria—Jos Plateau. 5. Agriculture—Social aspects—Nigeria—Jos Plateau. 6. Agricultural ecology—Nigeria—Jos Plateau. 7. Jos Plateau (Nigeria)—Social conditions. I. Title. II. Series.
DT515.45.K64S76 1996 96-9997
338.1'09669'5—dc20 CIP

British Library Cataloguing-in-Publication Data
A catalogue record for this book is available from the British Library.

For Bob

Na'an waar goegoe chagap.

Contents

List of Illustrations, ix

List of Tables, xi

Acknowledgments, xiii

1. Introduction, 3
2. Causality in Agrarian Settlement Systems, 12
3. Agrarian Production and Settlement, 28
4. The Kofyar Homeland, 57
5. Frontiering, 74
6. Pioneer Agrarian Settlement, 88
7. Land Pressure and Intensification, 101
8. Intensification, Dispersal, and Agglomeration, 118
9. Agricultural Movement, 129
10. Ethnicity and Settlement, 140
11. Settlement and the Physical Landscape, 159
12. Agrarian Ecology and Culture, 181

Appendix: Methods of Studying Agrarian Settlement, 197

Notes, 201

References, 215

Index, 251

Illustrations

3.1 Intensification slopes, 36
3.2 Development of settlement fixation, 46
3.3 Festive labor party in Ungwa Kofyar, 55

4.1 The southern Jos Plateau and the Benue Lowlands, 58
4.2 The central Kofyar homeland, 61
4.3 Marriage patterns between homeland villages, 70
4.4 Cluster analysis of marriage patterns in the homeland, 72

5.1 Drainage and soil zones in the Benue frontier, 75

6.1 Settlement and land use on the early frontier, 91
6.2 Ungwa in the core area, 1985, 98

7.1 Intensive farming in Ungwa Kofyar, 106
7.2 Social organization of agricultural labor inputs, all crops, 108
7.3 Dancing after a large mar muos in KDG Zang, 110
7.4 Social organization of agricultural labor inputs, selected crops, 112
7.5 Large mar muos in Dadin Kowa, 113

8.1 Example of frontier farm layout, 120
8.2 Length-to-width ratio and travel costs, 122
8.3 Farm shapes and length-to-width ratios, labor study households, 123
8.4 Nearest-neighbor distances in the core area in 1963 and 1985, 124

8.5 Typical compound and the Company macrocompound in the homeland village of Bong, 126

9.1 Detail of settlement in Goejak and Ungwa Kofyar, 130
9.2 Cumulative percentage of trips plotted against distance from compound, 132
9.3 Aggregate agricultural movement in Ungwa Kofyar, 133
9.4 Aggregate agricultural movement in Kwallala, 135
9.5 Agricultural trips across ungwa boundaries, 136
9.6 Drop-off of agricultural movement, 137

10.1 Microethnic composition of early frontier core-area ungwa, 1963, 145
10.2 Microethnic composition of frontier core-area ungwa, 1984, 147
10.3 Toenglu and work organization in Ungwa Kofyar, 1984, 153
10.4 Microethnicity of household heads in the core area, 1984, 156
10.5 Settlement encystment in the core area, 1984, 157
10.6 Settlement of Mwahavul farmers in Koprume, 158

11.1 Distances to nearest water in the core area, 161
11.2 Abandonment profiles, 167
11.3 Abandonment profile, piedmont zone, 168
11.4 Abandonment profile, Namu Sand Plains, 169
11.5 Abandonment profile, Jangwa Clay Plains, 172
11.6 Abandonment profile, southern Kwande area, 173
11.7 Rates of travel to local markets, 178

Tables

4.1 Household and Compound Sizes in the Kofyar Homeland, 64
4.2 Sargwat Endogamy, 71

5.1 Characteristics of Land Systems in the Study Area, 78

7.1 Populations for Core-Area Ungwa, 103
7.2 Labor, Movement, and Production in Kofyar and Kwallala in Labor-Sample Households, 116

8.1 Size and Population of Core-Area Farmsteads, 121

10.1 Makeup of 1984 Ungwa by Sargwat of Household Head, 148
10.2 Village and Sargwat Affiliations of Nearest Neighbors, 154

11.1 Farm Abandonment Rates by Soil Zone, 170

Acknowledgments

Part of chapter 12 of this book appeared, in slightly different form, in Stone 1993a. Some of the material in chapters 9 and 10 appeared in Stone 1991a and Stone 1992, respectively, although in both cases the material has been revised and developed. Passages of this book appeared in my doctoral thesis (G. D. Stone 1988), but overall the resemblance is slight. Readers interested in a more consciously archaeological treatment of this topic are referred to that work.

This book draws on work I have been doing since the early 1980s, and it has benefited from input from more quarters than I could list here. But there are a few who must be mentioned. Among the archaeologists, geographers, and sociocultural anthropologists whose input, insights, comments, and assistance have been helpful are Mike Adler, Jeff Bentley, Catherine Besteman, Tony Binns, Brian Byrd, Terry D'Altroy, Olivier de Montmollin, John Douglas, Chris Downum, Terry Jordan, Carol Kramer, Ken Kvamme, Gordon Mulligan, Anne Pyburn, Nan Rothschild, Mike Schiffer, Marilyn Silberfein, Rick Wilk, and Rich Wilshusen. I am grateful to Bill Doolittle, Tim Kohler, and Ben Orlove for their thoughtful comments on an earlier draft of the book, even if I didn't always follow their suggestions. In Santa Fe I gained much from discussions with Winifred Creamer, Dick Fox, Jonathan Haas, Barbara King, David Montejano, Doug Schwartz, and Peggy Trawick.

Among my students and assistants at the University of Arizona, Columbia University, and Washington University who deserve special thanks are Liz Bates, Kirk Dombrowski, Steve Ferzacca, Chris Kyle, Meredith Safran, Carla Singer, Daniella Soleri, Dee Williams, and especially Mark Sharifi.

Before and after fieldwork for this book I enjoyed hospitality in London from Barrey and Pris Sharpe, in Cambridge from Colin and Jane Renfrew, and in Durham from W.T.W. Morgan. Others who helped in England were A. T. Grove and Roger Blench at Cambridge and Rick Stone in London. In Kano, Mike Mortimore's help and wisdom were invaluable; for help in Jos, I am indebted to Hyacinth Ajaegbu, James Daduut, John Daduut, Gilbert Okechuku, and Rick Shain; for help from Lagos, thanks to Priscilla Keswani. Among the many Kofyar who helped in the field are Godfrey Daboer, Emmanuel Damiel, Alphonsus Dashe, Michael Dashe, Emmanuel Datson, Daniel Dayok, and the Hon. Lazarus Dakyen. I am grateful to Chris Downum, M.T.H. Glassgold, and Gordon Mulligan for sending me materials during my fieldwork.

For the use of the mirror stereoscope, I wish to thank Mike Schiffer and the Laboratory of Traditional Technology at the University of Arizona; for technical advice on photography, I am grateful to Greg Schmitz at Columbia University.

I must express my heartfelt gratitude to Chris Szuter of the University of Arizona Press. It was she who oversaw the review of the manuscript during times that were trying for us all, and this she did with professionalism and grace.

The Wenner-Gren Foundation for Anthropological Research provided support for my fieldwork in 1984–85 and also a Hunt Fellowship in 1993 to support my writing. The National Science Foundation supported the project through a grant to Robert Netting and dissertation grants to Priscilla Stone and to me. Additional funding came from the Educational Fund for Archaeology at the University of Arizona and the Council for Research in the Humanities and Social Sciences at Columbia University. A Weatherhead Fellowship at the School of American Research in Santa Fe provided a quiet year among the pines to analyze and write. I am deeply grateful for all of these sources of support.

Margaret Priscilla Stone, my long-term collaborator in much of the Kofyar research (and in some other significant endeavors as well), has commented on and contributed to all of the work incorporated in this book and just about everything else I've been up to for the past 15 years. Now that this book is done, I'll start cooking dinners again, I promise.

The last and hardest acknowledgment is to Robert McCorkle Netting, who began as my teacher and became my collaborator and friend. He was the true Kofyar scholar. In my first image of him he is lecturing on Kofyar farming; in my last, he is waving goodbye at JFK airport, a knapsack full of new Kofyar censuses slung over his shoulder. In between, he collaborated with me in two field projects and in numerous publications, commented insightfully on all my work, shared good sushi and bad puns with me, and helped in more ways than I could ever list. Throughout our association he showed me a standard of personal and intellectual integrity that I can only struggle to approach. Bob passed away in February 1995.

Settlement Ecology

1

Introduction

Namu is a surprisingly cosmopolitan town given its small size and its location away from major Nigerian travel routes. In its markets and alleyways one hears a dozen local languages, as well as Tiv, Igbo, Fulbe (Fulani), Hausa, and pidgin English. It is an agricultural boom town, located on the northern edge of what has been transformed in the past few decades from a vast stretch of lightly inhabited savanna to a highly productive and heavily settled breadbasket area for the burgeoning Nigerian cities. The agricultural boom has drawn farmers from all directions, along with traders to handle the produce, and diverse entrepreneurs.

Traveling the dirt road south from Namu, one encounters these farmers very quickly. After driving over a small dam and past the perimeter cultivated by Namuites one reaches a place called Ungwa Long. *Ungwa* is a Hausa word meaning a settlement district or ward; *long* means chief in the language of the Kofyar, the people whose home was a day's walk to the north on the Jos Plateau. From there, dirt roads and paths lead to Goewan, Dunglong, and Pangkurum, all names taken from Kofyar villages in their homeland to the north. The Kofyars' movement into this frontier south of Namu began in the 1950s. At first the farms were seasonal outposts, but they gradually became homes as the plateau was abandoned.

In some ways, the frontier settlements resemble the villages for which they are named; most of the population is dispersed in adobe compounds on contiguous farms, each owned and worked by a single household with assistance from neighbors. As in the homeland, groups of adjacent farms form *toenglu* or neighborhoods, which are units for labor pooling; several toenglu constitute a village or ungwa, which is a larger labor pool and also a political entity.

At the same time, frontier settlement patterns were markedly different from the original villages. In some spots, compounds clumped together into hamlets. Frontier farms were much larger, and the residences often incorporated new architectural forms. Frontier settlement patterns have a linear quality, with residences attracted to roads and paths in a way that homeland compounds never were. The arrangement of frontier farms with respect to microethnic affiliations had been shuffled and redealt. Settlement stability had also changed: homeland compounds were enduring, their locational stability commemorated by cairns marking burials of generation after generation of household head; on the frontier, farms were often relocated or abandoned, sometimes after only a few years. Patterns of production and settlement had also changed through time and with the change in location, with cultivation becoming more intensive, land tenure becoming more private, residential clustering declining, attraction to water declining, and attraction to high-quality soil increasing. In forging this landscape, the Kofyars' decisions were unaffected by development agencies or government.

I first heard of the Kofyar as a graduate student in archaeology, a discipline long preoccupied with settlement patterns. Yet the accumulated wisdom of archaeology—and even of geography, the field manifestly devoted to the spatial—was surprisingly silent on how agrarian settlement patterns were governed, and there was no body of theory that could have predicted or explained the Kofyar landscape. What were the underlying principles of this landscape?

Dorf and *Stadt*

Christaller's *Central Places in Southern Germany* begins by asking whether there are laws that determine the number, sizes, and distribution of towns. Before presenting his famous theory of central places, Christaller makes it clear that his concern is with the town (*Stadt*) and not the village (*Dorf*). In the first paragraph of the book he lays to rest the *Dorf* question:

> The root of the village [Dorf] is distinct: it is typically agriculture and other land uses. The connection between the number of people living in villages and on farms, and the size of land area, is stated in the following manner:

> There are as many people in a given area as can live from the cultivation of the land with given agricultural technology and organization. Whether these people live in large settlements, i.e., large closed villages, or in smaller villages, hamlets, or on individual farms is not clear *a priori*. However, the investigations of Gradmann and others have clarified the issue, for they say that a particular type of settlement is usually predominant within a certain region (Christaller 1966:1).

My book too asks how the number, sizes, and distribution of settlements are determined, but I am interested in the *Dörfern*, which Christaller dismissed as a matter of cultural tradition. For Dorf I use the term *agrarian settlement*. These settlements are characteristically inhabited by a few to a few hundred persons (mostly or exclusively farmers); the settlements as defined include discrete residential compounds, however small, and exclude communities with a significant proportion of non–food producers. Like many concepts in social science, this lacks the precision that definitions enjoy in mathematics and physics, but it is useful (in fact, essential) nonetheless.

Christaller's opening has troubling implications from the perspective of anthropological archaeology, a field that has spent the past 30 years wrestling with issues of nomothetic explanation and 40 years focusing on settlement patterns, most of which have been agrarian settlement patterns.[1] Settlement pattern analysis has been central to anthropological archaeology since Willey's work in Peru's Virú Valley (Willey 1953), and it had become pivotal in American archaeology by the 1960s, fueled by the new archaeology's interest in regional systems of adaptation (Binford 1964; Struever 1968). By the early 1990s, more archaeology symposia dealt with settlement and regional systems than any other topic (Feinman et al. 1992).

Yet we have a wide and growing disparity in settlement theory, with our understanding of hunter-gatherer settlement far outstripping what we know about agrarian settlement. The foundation for hunter-gatherer settlement theory has come from research on hunter-gatherer ecology, but models of the organization and evolution of agrarian settlement have been curiously oblivious to agrarian ecology. The approach in this book is to look explicitly at agrarian settlements qua farming communities, an approach I call settlement ecology.

The archaeological study of settlement pattern has always had strong ties to ecology—indeed, it began as an idea of Julian Steward's (see Willey 1974:153, 1953:xviii), and in the settlement studies that flourished in the years following Willey's 1953 monograph, it was apparent that "the concept of settlement pattern should include the ecological dimension at the outset and thereby provide the basis for the other . . . types of analysis" (Vogt 1956:175). Ironically, it was with the new archaeologists of the 1960s, a fundamentally ecological crowd, that research on agrarian settlement began to overlook the role of food production. Although classic studies of hunter-gatherers dealt with relationships between subsistence and settlement (e.g., Struever 1968; Winters 1968; Binford 1968b, 1980), agrarian societies served more as laboratories for the investigation of ancient social organization (e.g., Deetz 1968; Whallon 1968; Longacre 1970; J. Hill 1970; cf. Ashmore 1981:38–39).

Exploration of the relationships between settlement and hunter-gatherer ecology has continued (e.g., Binford 1980; Kelley 1985; Thomas 1972, 1988), often becoming unabashedly environmental-determinist (Binford 1990). Agrarian settlement, by contrast, has been harder to link to environment, and progress in understanding the dynamics of agrarian settlement has been frustratingly slow. Indeed, we are now seeing studies that strain to apply explicitly hunter-gatherer models to agrarian settlement (e.g., Hard and Merrill 1992).

Pattern and System

It is not even clear what an agrarian settlement theory should look like; should it comprise formulae, general principles, or specific rules? In archaeology, one answer involves the concept of settlement *systems*, defined as underlying sets of "rules" responsible for the empirical settlement *pattern*. The idea of settlement rules had been tried out before in geographical studies, some of which linked the rules to other theories of human behavior (e.g., Aldskogius 1969) and some of which treated settlements more as simple points on a surface (Bylund 1960).

The archaeological version was illustrated by Flannery (1976a:180), who proposed a set of rules that would have produced the Formative period settlements in the Etla Valley of Oaxaca. To summarize the "settlement system," (1) settlement begins near a good ford on the river; (2) it begins spreading symmetrically, with daughter settlements locating midway between parent settlements and valley margins; (3) later, daughter settlements are located midway between parent settlements and other settlements; and (4) infilling stops at a socially determined spacing, but satellite settlements may continue to form at special resources. The settlement pattern was analogous to a phenotype, whereas the "rules" were the genotype. Because the example was meant more as an illustration of settlement system rules than as a reconstruction of the actual Etla Formative system, there was no attempt to explain the genotype itself, that is, why Oaxacans might have followed this, as opposed to another, set of rules.[2]

This raises a host of questions that are vital to an understanding of agrarian settlement. For instance,

> Why would daughter settlements in Oaxaca *maximize* distance from parent settlements, rather than *minimize* this distance as assumed in other models?[3]
>
> Agriculture often becomes intensified when population density rises (chapter 3); why then would Oaxacan farmers move to new sites instead of staying and intensifying their agriculture?
>
> In Flannery's system, new site loci are a simple function of old site loci; when should we expect landscape features to override those social factors?[4]
>
> In a more theoretical vein, one wonders where "rules" of settlement systems come from. In what ways are they historically contingent, and what are the cross-cultural regularities?

But even more important than these questions—in fact, at the heart of agrarian settlement theory—is the problem of equifinality (Bunge 1962; Crumley 1979; Hodder and Orton 1976): the number of sets of rules that could have produced the settlement pattern is a large number indeed, limited by little beyond the investigator's imagination. If we are going to look for rules of settlement, how can we identify which rules any agricultural population followed? It seems clear to me that any proposed settlement rules (whether for minimizing or maximizing distance between settlements, for seeking proximity to rivers or to large plots or to rich soil, or for what conditions cause farm abandonment) must be informed by some other information about human behavior.

In a more theoretical vein yet, we have to review the concept of rules and of determination in general. Rules are not only a conceptual tool for explaining settlement patterns; they are also essential for the simulations used to evaluate settlement models (e.g., Bylund 1960; Aldskogius 1969; White 1977; Keegan and Machlachlan 1989). But rules, as the maxim goes, are made to be broken, and—barring the trivial—any settlement rule can potentially be overridden. Flannery was aware of this, but his response was to admit his rules to be probabilistic (containing a random element) rather than deterministic (Flannery 1976b:171). In effect, deviations from the rules are simply noise.

However, settlement decisions can be affected by myriad different factors, and it seems to me that these decisions must be guided by a mental balance sheet for each factor. To propose a rule is to claim one factor to be overriding in a particular situation, yet rules vary in how overridable they are, and by what. Rather than single rules for site spacing, abandonment, and so on, I prefer to think of priorities of varying strength. The ease with which these are overridden and the types of factors that tend to override them are not noise but a central aspect of settlement study (e.g., Farriss 1978; Silberfein 1989; Galt 1991).

In sum, Flannery provides a starting point for thinking out what a theory of agrarian settlement should look like, but settlement rules cannot exist sui generis, and they should be neither deterministic nor probabilistic. We should begin by considering agrarian sites as components in a farming system, seeking to understand the premium this places on different settlement decisions, and exploring how this premium interacts with, and can be overridden by, other factors.

Predicting and Retrodicting Settlement Patterns

When first studied by Robert Netting in the 1960s, the Kofyar were colonizing a sprawling agricultural frontier in the Benue Valley, leaving their crowded homeland in the rugged hills of the Jos Plateau. Netting had described the settlement pattern of contiguous farmsteads in the home area, but the frontier settlements had not been mapped, and aerial photographs were not available.

The unknowns about Kofyar frontier settlement and its evolution since the 1960s posed a vital question: in theory, what settlement pattern should we expect? Given a group of paleotechnic farmers moving into a frontier, how would we begin to model the size, shape, and duration of individual sites and their spacing with respect to each other and landscape features?

The null hypothesis, that frontier settlement would replicate the homeland pattern, was out of the question; the differences in landscape were too great, but more important, there was no theoretical basis for predicting whether the homeland pattern *should* be replicated. Despite decades of studies by geographers and anthropologists comparing settlement patterns before and after migrations,[5] there was neither consensus nor theoretical justification for explaining how premigration settlement should influence postmigration settlement.

Surely the Kofyar frontier settlements would adjust to the physical landscape, their locations being affected in some way by water and arable land. But again, archaeological and geographical studies of the effects of such resources on settlement collectively failed to furnish a set of expectations for Kofyar settlement.[6] Would settlers choose proximity to water over land quality? Would these priorities remain constant, or might they change with population pressure? Would distance to either resource be minimized, or merely kept below a certain threshold? Were such distance thresholds dependent on population density? And to what extent would such locational "rules" be dictated by the farming system?

While these questions deal with how people relate to their landscape, there was a set of knottier questions about how people relate to each other as they go about the business of farming. What would the social units of production be, and how would they affect settlement organization? Would the population be atomized, aggregated, or something in between? If aggregated, would the aggregations comprise autonomous economic units? (And if so, why?) If dispersed into household settlements, would there be any suprahousehold labor organization? (And if so, why?)

Despite the wealth of research in several disciplines and on a wide variety of rural settlement patterns, there was no basis for predicting the settlement pattern I would see in Nigeria.

Searching for Causes

In archaeology we have been more inclined to enumerate influences on settlements than to ask which factors were instrumental and why. Trigger's (1968) paper on the determinants of settlement location notes the importance of subsistence regime, availability of building materials, environment, family structure, wealth differentials, specialization of production, political institutions, secular tastes, subsistence technology, migration, and population; Sanders (1981) offers a more ecological list of determinants, including rainfall, temperature, topography, hydrography, and zonal soil patterning, as do Brown et al. (1978:169). But these lists do not tell us which variables should take precedence in any given situation, or when variables are "satisficed" (exceeding a threshold) rather than optimized (Hamond 1981:223). The most successful locational models, such as Christaller's theory of town settlement and von Thünen's theory of land use, are grounded in well-established premises concerning profit maximization and transport cost; but we have had difficulty grounding rules of agrarian settlement in any theoretical matrix. In archaeology we have tended more toward empirical generalizations than rules backed by theory.

Other disciplines to which we have turned for inspiration have not had exactly what we need. Cultural anthropologists have shown only a fleeting interest in agrarian settlement patterns, focusing mainly on markets (Skinner 1964, 1977; C. Smith 1975). The spatial purview of ethnography is generally too small for settlement pattern analysis; as Geertz points out, ethnographers do not study villages, they study *in* villages.

The geographers, whose help on issues of market settlements has been immeasurable, have been of less help with agrarian settlement rules. Geographical descriptions and classifications of agrarian settlement have often been splendid, especially for Europe (DeMangeon 1927, Mayhew 1973; Roberts 1977), but settlement rules have generally not been dynamic (allowing change through time) or robust (applicable to other settings). Archaeologists have filled the need for a theoretical model of agrarian settlement evolution by turning to a paper by the geographer Hudson (1969), but his model has turned out to be inappropriate and broadly unsuccessful. Although purporting to model settlements under rising population, the model is oblivious to the agricultural changes that occur with population pressure, which have profound effects on settlement. Taking a diametrically opposite tack, my explanation of agrarian settlement evolution begins with the conditions of agricultural change.

Production and Settlement

Ironically, Hudson's theory appeared just as the full impact of Boserup's *Conditions of Agricultural Growth* (1965) was being felt and ethnologists, agrarian geographers, agricultural economists, and archaeologists were being alerted to the relationship between population and agricultural change.

The impact of Boserup's work (1965, 1981) on agrarian research has been profound, and it would be impossible to make sense of Kofyar settlement without it. Yet hers is not a model of settlement pattern. Like von Thünen's (1966 [1826]) famous theory of land use, it holds constant or ignores settlement pattern in order to isolate a fundamental relationship between population and agricultural production. That is the starting point of the present book. I will use an unusually rich set of data on agriculture and settlement, collected among Kofyar farmers who were in the process of filling a frontier in the Nigerian savanna, to examine the social and spatial entailments of intensification.

Although this is a study of a particular group of people in a particular place and time, my ultimate interest is in general relationships underlying cultural behavior. We cannot make sense of variability in rural settlement without considering the effects of culture and history, but we cannot even begin to make sense of settlement variability without an understanding of this ecological substrate. My focus may be agrarian settlement ecology, but agrarian settlement patterns cannot be appreciated as simple artifacts of ecology. The goal in this book is a better understanding of the linkages, both in theory and in my empirical study in the Nigerian savanna, between the social and the spatial organization of production.

2

Causality in Agrarian Settlement Systems

Our success at explaining ancient and modern settlement systems alike hinges on how well we understand the effects and relative importance of different causal factors—what causes determine which aspects of settlement pattern, and why. Let us begin by considering what the fundamental determinants of agrarian settlement may be. What we ultimately want to understand is how and why settlement patterns change through time, which means understanding how the relationship among causal agents evolves. We know less about this than one might imagine, but a selective and thematic tour through previous work will be valuable.

This is not a synopsis of the literature on agrarian settlement patterns per se; that literature is too broad to lend itself to cogent generalizations or meaningful overviews in a work of this scope. Reviews focusing either on the chronology of ideas or on some aspect of settlement archaeology may be found in Trigger 1968, Parsons 1972, Hodder and Orton 1976, G. Johnson 1977, Roper 1979, Crumley 1979, Evans and Gould 1982, and Vogt 1983. Summaries of the geographical literature are also numerous; interested readers may want to consult Baker 1969, Chisholm 1979, Eidt 1984, and Jordan 1966.

Instead, my focus is on causal arguments in agrarian settlement models. I believe that some of the most widely used approaches to agrarian settlement patterns are based on inappropriate premises. Particularly lacking has been the integration into settlement research of those principles of agrarian ecology that have been elucidated in recent years.

The factors affecting the size, location, and duration of agrarian settlement are obviously many; explaining systems of settlement is more difficult still. We need a body of theory dealing not only with the effects of various factors, but with their relative importance. When the farmer wants to be on fertile soil, near water, and adjacent to a brother who will provide help, where do the priorities lie?

What are the factors that push and pull agrarian settlements? The list is as long as one cares to make it (Trigger 1968). As Isard (1960:3) notes, settlement patterns are rife with problems of mutual causation, but we must "cut the circumference somewhere." The following discussion reviews studies that best represent the various ways the circumference has been cut.

Residence and Production: Von Thünen and Chisholm

Von Thünen's (1966 [1826]) classic model of rural land use does not deal with determinants of settlement pattern; in fact, it establishes an artificially simple pattern of settlement as one of its assumptions. But it is fundamental to this discussion because in modeling how land-use patterns are determined by spatial relationships to settlements, it builds a theory important to the determination of settlement location. The extension of this body of theory to settlement location is one of the major thrusts of Chisholm's *Rural Settlement and Land Use* (1979).

The von Thünen model is important because it was the original attempt to isolate the effects of one variable on the spatial economy. His main contribution was not the relationship between proximity to town and land use; that notion went back at least as far as Adam Smith's (1776) *Wealth of Nations*. Rather, his originality lay "in his method of partial equilibrium analysis, or putting most factors at rest" (W. B. Morgan 1973:301). As Chisholm put it, "von Thünen himself was at pains to point out his particular findings had no claim to universality. But, as he rightly noted, the analytical method he employed could be applied generally . . . it is the method and not the particular finding that counts: . . . an essay in *a priori* reasoning" (1979:14). The model establishes the land-use pattern that should result from a few fundamental and ubiquitous factors, and then considers how variability in other factors should affect the baseline pattern.

The model in von Thünen's *Isolated State*, and Chisholm's elaboration on it, has been summarized too many times (e.g., Haggett 1965; Henshall 1967; Abler et al. 1977) to warrant extensive treatment here. It predicts land-use patterning based on economic rent, defined as "the surplus produced by the application of labour and other inputs at a given site compared with the return obtainable from the cultivated land the most remote from the central city" (Chisholm 1979:17). Economic rent is comparable to marginal productivity in two senses: it refers to the relative returns of cultivating different crops in different locations, and in the von Thünen model it is applied to expanding the margin of cultivation.

The heart of the model is that maximization of economic rent in a theoretical "isolated state"—a homogeneous agricultural region with one market town—produces a land-use pattern of concentric rings, because transport costs increase and profit margins decrease with distance from the town. This is an expression of a simple principle of unparalleled importance in settlement theory, which I will call the proximity-access principle. It simply states that the greater one's need to access any landscape feature, the greater the premium on residing near that feature.

The engine in von Thünen's use of the principle is profit maximization, which in von Thünen's scenario affects land use through the expense of transporting agricultural goods. The elegance of the model comes from its unification of several factors within the single framework of monetary profit.

However, models of farmers producing little or nothing for the market must not only abandon monetary profit as a variable to be maximized but must also rethink the role of the maximization concept in general. In translating the von Thünen model into terms relevant to subsistence economies, Chisholm sought cross-cultural comparability by turning to the common currency of time; considerations of maximum cash return were translated into principles of minimizing traveling distances and labor inputs. The application of the concentric zone model to individual farms had originated with von Thünen, who had pointed out that time management on individual farms should replicate patterns produced by profit management in the region, leading to decreasing expenditure of labor as distance from the residence increases. At some point cultivation ceases to be profitable; therefore there is an optimum size of farm for any system of production and a limiting distance beyond which cultivation is disadvantageous. However, this argument was not developed and had little influence until the following century, when it was taken up by Chisholm.

Chisholm had access to a wealth of studies from agrarian geographers revealing how the von Thünen concentric zone model was manifested in subsistence economies and in particular how the principle affected the operation of individual farms. Studies worldwide have confirmed the prevalence of concentric land-use patterns (e.g., Prothero 1957; Henshall 1967; Horvath 1969; Haggett 1972) or other patterns that may be derived from von Thünen's argument (e.g., Sallade and Braun 1982). Chisholm stresses the broad pattern of agricultural activity being concentrated within a radius of 1–2 km from settlements, activities beyond this distance being strongly affected by travel time and generally terminating between 3 and 11 km.

The von Thünen model and Chisholm's reworking of it are particularly interesting taken together in that they suggest a principle of land use that crosscuts commercial and subsistence economies. To scholars interested in causality in settlement, von Thünen's main contribution is the modeling of the effects of management of transport costs, whereas Chisholm's main contribution is the demonstration that similar patterns are produced by the management of time. Following Chisholm's work, Carlstein (1982) argues that time, the principal scarce resource in preindustrial economies, is economized as money is in market societies.

The von Thünen model has had an important if indirect impact on archaeological studies of settlement. The principle that he stated and Chisholm elaborated led to a set of expectations concerning villages and their agricultural radii. This was the inspiration for site catchment analysis in archaeology—the interpretation of site locations by analysis of resources available within given distances. Through catchment analysis (or more properly, territorial analysis), the von Thünen/Chisholm model has influenced the investigation of hunting-and-gathering settlement (Vita-Finzi and Higgs 1970) as well as agrarian settlement (Ellison and Harriss 1972; Flannery 1976a, 1976b; Zarky 1976; Rossman 1976). The relationship between the von Thünen/Chisholm model and territorial analysis is explored below.

Resource Exploitation

Chisholm inverts his analysis to treat settlement pattern as the dependent variable. Proposing that settlements will locate themselves so as to minimize transport costs—or travel time—raises the issue of how settlers weight various resources and how frequently they access those resources. In his hypothetical illustration, Chisholm assigns attraction weightings to landscape features: water (10), arable land (5), grazing land (3), fuel (3), and building materials (1).

How strongly landscape features attract settlements under actual conditions is not known. Other studies have cited landscape features without attempting to justify a weighting; for instance, Price and Price's study in the Ozarks (1981) predicts agricultural pioneer settlement to be pulled toward water, firewood, and productive farmland (as well as trade and communication routes) without explaining the relative attraction of these features. In Chisholm's example, the high value of water reflects the fact that "it has to be used at frequent intervals in the day and is difficult to carry and store in large quantities when only elementary implements are available, such as pitchers and gourds" (1979:95). Chisholm regards as atypical (1979:102) those cases where the attraction value of water is relatively low, such as the settlement patterns of the Ngwa (W. B. Morgan 1955a) and others in eastern Nigeria (Karmon 1966), where residences are located up to 13 km from water. However, in chapter 11 I demonstrate how the attraction value of water can be expected to vary with land pressure, decreasing as farmland becomes scarce—a pattern exemplified quite neatly in Morgan's data.

Territorial (Catchment) Analysis

The general problem of the attraction of settlements to landscape features is one for which archaeology has tailored the methodology of site catchment analysis, which is better termed territorial analysis.[1] This method attempts to identify the effects of landscape features on settlement location by examining the contents within specified radii from the site. The method, developed by Vita-Finzi and Higgs, has been demonstrated in studies of paleolithic settlements in Greece (Higgs et al. 1967) and Natufian settlement in Palestine (Vita-Finzi and Higgs 1970).

Although the method was developed for analysis of hunter-gatherer settlement, it is applicable to agricultural systems. Roper's statement of the underlying theory underscores its close relation to the von Thünen and Chisholm models: "It is assumed that, in general, the farther one moves from an inhabited locus, the greater the amount of energy that must be expended for procurement of resources. Therefore, as one moves from that locus, it is assumed that the intensity of exploitation of the surrounding territory decreases, eventually reaching a point beyond which exploitation is unprofitable" (1979:120). The problem in territorial analysis is the determination of catchment radius (variation in territory sizes is summarized in Stone 1991a). The question of how far the "reach" of various environmental features extends will have different answers for hunter-gatherers and farmers, and there are important distinctions to be made between different types of hunter-gatherer strategies.

However, my concern is with farmers, and there is a considerable amount of material relating to catchments, from both theoretical and empirical viewpoints. Chisholm's survey leads him to posit a threshold of approximately 1 km, beyond which residence-to-plot distance plays a major role in determining intensity of labor input. Limiting distance (the distance beyond which farmers are unwilling to travel on a frequent basis) is suggested to range from 3 km to 11 km depending on size of settlement and climate (Chisholm 1979:61).

Territory size turns out to be a crucial variable in settlement decision making, as we will see in the following study (see especially chapter 9). Here I stress that territorial analysis is not a theory of site location but a method of investigating the relationship between site location and landscape features. As it has usually been practiced, it is based less on the von Thünen/Chisholm theory of land use than on the corollary that the farther a feature is from a site, the less it affected the locating of that site. I would like to move beyond the simple formulation that the landscape's effect on settlement decays with distance. With the Kofyar I look at the relative effects of different landscape features on settlement location, and, more important, how the relative effects change with agricultural intensification (chapter 11).

Marketing Behavior: Central Place Theory

As developed by Christaller (1966), central place theory is a partial equilibrium analysis of the effects of the marketing of goods and services on the spatial arrangement of towns. It begins with the concept of central functions (goods and services), which "are produced and offered at a few necessarily central points in order to be consumed at many scattered points" (Christaller 1966:19). Central functions with increasingly high thresholds (the minimum market required to bring into being and support that function) and increasingly long ranges (the distance consumers will travel to buy the good or service) can be offered only in increasingly widely spaced towns (Christaller 1966:22; Abler et al. 1977:364–371).

The theory states that if we hold constant terrain, transport facilities, and the distribution of population and purchasing power, towns offering central functions should be spaced equidistantly, forming a hexagonal lattice pattern (Christaller 1966:63). The spacing of towns offering functions with increasingly high thresholds—higher order functions—is increasingly distant, leading to a pattern of nested hierarchical lattices (Christaller 1966:66). This is an application of the proximity-access principle to secondary (i.e., nonagricultural) features on the landscape.

Christaller refers to the prime mover in the theory as the market principle, which is based on "the range of the central goods, from the point of view that all parts of the region are supplied with all conceivable goods from the minimum possible number of functioning central places" (1966:72). He also recognizes the traffic principle, which seeks to maximize the movement of goods while minimizing the transport costs, and the sociopolitical separation principle, which allows for central places to develop for defensive, administrative, or other reasons beyond the marketing of goods (1966:74–80).

Because Christaller's theory is at its core a partial equilibrium analysis of the marketing principle, he does not develop the argument of the effects of administrative functions on settlement. His causal argument is specific to market towns, and Christaller was clear in stating that agricultural settlement patterns were determined separately. The later refinements on the Christaller model by Lösch (1954) delved much further into the economic aspects of location. As an analysis of spatial equilibrium in a single industry (a brewery), Lösch's model is more intertwined with theories of consumer and firm behavior than was Christaller's formulation.

Central place theory has undergone modifications since its appearance (e.g., Olsson 1966; Marshall 1969), but it remains essentially a theory of the location of towns and cities and therefore is beyond the scope of this book. But a brief consideration of archaeological uses of central place theory will be instructive as to causality in settlement systems in general, especially because archaeologists have been concerned with the relative importance of the effects of central functions and site territories on settlement. For reviews of anthropological, and especially archaeological, applications of central place theory, see Crumley 1979, Evans and Gould 1982, and King 1984.

Whereas geographers, true to the original basis of the theory, have actively studied central place patterns from an economic point of view (e.g., King 1962; Berry and Barnum 1962), archaeologists rarely have sufficient data on the goods and services offered by prehistoric settlements to adopt this perspective (Crumley 1979:153; see also Adams 1974). One approach has therefore been to compare observed settlement patterns with the idealized lattice model in order to make inferences about the economic functioning of sites (Hodder and Hassall 1971). Another use of the model has been to compare observed settlement patterns with the lattice model to show the existence of the theory's boundary conditions. For example, an analysis of Aztec settlement marketing system concludes that commercial factors were paramount in shaping the pattern, albeit moderated by agricultural considerations (M. Smith 1979:121; see also Evans 1980).

Christaller's theory was not, as some have claimed, an entirely static model (see Preston 1985; Smith 1974), and archaeologists have linked it with theories of cultural evolution. The causal argument is that as societies come to be integrated at higher levels of complexity, towns take on administrative central functions that come to dominate ecological factors in settlement (Flannery 1976b:170; G. Johnson 1972). Earle (1976:219) develops this idea within a neo-evolutionary framework, relating a single-tier settlement hierarchy in the Basin of Mexico to a tribal level of organization, and a two-tiered settlement hierarchy with a chiefdom level. Steponaitis (1981) presents a model in which the growth of settlements beyond a size consistent with their catchment productivity is linked with flow of tribute and thus with political evolution.

Fissioning and Spacing

Siddle (1970) provided a model of settlement change resulting from population growth in a small subsistence village. Population in such a village is expected to grow until it cannot be supported within the village's economic radius, forcing a change in the equilibrium. Villagers reject reduction of fallow because of "the subsistence cultivator's inherent awareness of his environment" and reject improved farming methods because of "innate conservatism" (1970:80). Rather, the economic radius is extended by establishing temporary farm settlements, a process that leads to village fissioning.

Farmers are reluctant to fission because fissioning separates closely associated kinsmen, and so "the new settlement would be located as close to the old as the economic radii of both settlements would allow. . . . In the idealized economic landscape this would be at a distance of twice the normal 'economic' radius of the founder" (1970:80). The original village continues to fission while the clones remain static, until a hexagonal lattice develops. Settlement change is assumed to be driven by the desire to remain as close to kinsmen as the principle of economic radius allows.

Bylund (1960) presents a model of settlement colonization in which the location of each new settlement node is a function of the locations of (1) other settlements and (2) landscape features (road, church, and market). Each new settlement is a "clone" of its mother, and the primary force governing settlement pattern development is intersettlement distance: "It has been considered an important endeavor of the colonists to find and choose new land as near the mother settlement as possible" but also "to have the new settlement in a position as free as possible from land competition from other pioneers coming from other mother settlements" (Bylund 1960:45). The secondary force in Bylund's model is the array of attraction values: "The attraction of the settlers' lands is inversely proportional to the perpendicular distance from the road. . . . The church and market-place are assumed to exercise a certain attraction on the settlement" (Bylund 1960:45).

Bylund assigns attraction values based on these forces to a grid in an attempt to model the colonization of a test area in Sweden, with equivocal results. He finally introduces the notion that land selection criteria, especially regarding farm size, may change as a settlement pattern unfolds (for instance, when technological change reduces the minimum amount of land required).

Morrill's (1962) simulation is similar, although his concern is with the evolution of settlement patterns predicted by central place theory, and he uses no empirical data. The model is probabilistic and is based purely on settlement size and intersettlement distance. Settlement populations grow in mechanical increments of two, one of which is sent off to found a daughter settlement. The probability of the daughter's settling on a given square is proportional to that square's proximity to the mother settlement.

Morrill (1963b) has also provided a discussion of the statistical distributions of migration distances generated by gravity models (i.e., models in which the location of daughter settlements decreases proportionately with distance). Morrill's studies are in fact a sort of partial equilibrium analysis, trying "to understand fundamentally the role of distance and area in behavior" (1963b:76), but the factor whose effects are being evaluated is essentially a statistical formula. The premise of the gravity model is widely accepted (Neyman and Scott 1957; Hudson 1969:370; Bylund 1960) but clearly requires greater empirical examination. However, from the vantage point of causality, the question is why daughter settlements should be expected to locate close to the parent in various situations.

Archaeologists have used and modified these geographical fission models, in some cases providing their own causal arguments for the spacing of daughter settlements. The Bylund model appears in Hamond's (1981:215) account of new Linearbandkeramic (LBK) farmsteads being located as close to their "parent" as possible, on the best possible land, and on land most like the parent settlement's land—to minimize the cost of movement to the new site and to maximize the certainty of the new environment.

What constitutes land "most like" the parent settlement's land is an open question, and it is likely that the soil distinctions made by the LBK farmers were extremely coarse (Wilshusen and Stone 1990). And, although there are reasons why movement cost between sites might be minimized, it is not clear in this case why movement cost would be minimized. When farmers abandon one site for another, dependence on stored or scavengeable materials from the first site would promote low intersettlement distances. As examples, a major reason for short migration distances among the Venezuelan Wothiha is the need to fetch manioc cuttings from old fields (Zent 1992); Anasazis regularly scavenged beams from recently abandoned villages (Ahlstrom et al. 1991); Raramuri winter rockshelters are located relatively close to the summer dwellings, to which the Raramuri regularly return to fetch stored foods (Hard and Merrill 1992). But in the LBK case, it is unclear what the farmers needed to get from the still-occupied parent settlements. It was not labor, according to Hamond; he assumes (1981:222) that each site housed a completely independent labor pool.

In contrast, Flannery's (1976a) study of site spacing in the Etla Valley of Oaxaca proposes that agricultural villages may strive to keep intersettlement distances greater than the economic radius: new villages are posited to be located midway between previously founded settlements (1976a:180). Flannery stresses the discovery of "a set of 'rules' that generated the pattern in the first place" (1976b:162), but his set of San José settlement rules pay little heed to agricultural production; in fact, yield per hectare is taken to be a constant (1976a:177). The agricultural options that would actually have been available, and their role in determining site location, are not considered.

Models treating settlement location as a function of earlier settlement locations have an elegance that may account for their appearance in various lines of settlement research. However, as this discussion has suggested, they often require overly limited views of the interactions between nodes in the settlement system.

Social physics models are so named because of their explicit use of physics concepts, such as gravity, in characterizing relationships between distance and interaction. Distance and interaction could even be reduced to mathematical formulae, as shown by geographers in the 1960s (Haggett 1965); the 1970s saw numerous explorations of such models by archaeologists (e.g., Plog and Hill 1971; Wilmsen 1973; Plog 1976). But like much of the quantitative geography of that era, the strength of these techniques was in the characterization of relationships and not the explanation of *why* those relationships occurred. "Interaction" is a rubric that surely encompasses a range of behaviors with profoundly different effects on settlement spacing, and the 1980s brought calls for attention focusing "less on the intensity of interaction than on its nature" (Hodder 1981:92; Moore 1983:181). Put differently, gravity models of intersite interaction are consistent with the proximity-access principle, yet they are hollow without stipulation of how the sites are interacting and why. In this book I show how it is specifically interaction in agricultural production that affects site spacing, and how the relationship can be understood only by first understanding the social organization of agriculture.

Econiche and Competition: Hudson and Related Issues

Hudson's (1969) model of the evolution of agrarian settlement requires a close look, not only because of its wide influence but because it exemplifies some important and widespread assumptions about the social aspects of settlement. The model predicts three stages of settlement, the first of which states the general process of people moving into a new area in formal terms of niche and biotope. His example is of technological change opening up a new area for a certain kind of production: the railway reaching the Great Plains increased niche volume, promoting grain cultivation (1969:367). Colonization is summarized as a function of increase in the size of the fundamental niche. The second stage is a formal statement of population increase. Settlement clones tend to locate close to their source settlement, producing clusters, and/or exogenous entrants avoid preexisting settlements, producing dispersal. The third stage concerns the effects of density-dependent processes on spread. The density-dependent process is "the lower limit on the size of farm that can be operated economically" (1969:371). Larger settlements absorb smaller ones in a process likened to large trees outcompeting smaller ones in the process of succession. This process, coupled with farmers' desire for compact holdings to minimize travel costs, promotes the development of regularly spaced farms.

Hudson gives no boundary conditions for the theory. He was less concerned with causal mechanisms in settlement evolution than with a formal statement of the processes involved; that is, the translation of concepts such as migration and site selection into terms of biotope and niche space. In working out testable implications of the model, Hudson focuses on integrating his concept of interfarm competition with statistical probability to generate hypotheses regarding relationships between settlement density and clustering. The implications he tests may be paraphrased as follows: because settlement density is lowered by competition among fixed points, where reasonably high densities decline, the distribution will tend to increase in regularity; because settlement density increases in the second stage tend to be produced by the formation of daughter settlements close to the parent, increases in clustering should be seen mainly in cases where density has increased; and competition should be most in evidence where there is farm abandonment and an increase in farm size (1969:377).

The theory contains specific implications regarding settlement pattern only for the final stage, where the mechanism driving settlement evolution is competition forcing out inefficient farms. A considerable amount of substantive content has been read into the theory regarding its implications for settlement pattern, with settlement in the three stages being variously interpreted to be random to clustered to even (e.g., Warren and O'Brien 1984:39–42) or clustered to random to uniform (Haining 1982:212). Even the cornerstone concept of eventual even dispersion has been subjected to reinterpretation by geographers; Austin (1985:205) argues that "intensified stability in Hudson's third, competitive stage would seem to lead to villages" in central Sweden.

The theory has had enormous influence in archaeology. Both Wood (1971) and Hodder (1977) have discussed the application of the model to prehistoric settlement data; Stark and Young (1981:295) have used it in their discussion of settlement spacing. It has been applied to historic period settlements in Connecticut by Swedlund (1975), in northeastern Missouri by O'Brien (1984), and in South Carolina by Lewis (1984). Paynter (1982:115) has used it in modeling spatial relations among producers for agrarian markets, Blouet (1972) based his discussion of settlement evolution on it, and G. Johnson (1977:493) has discussed its implications for the characteristics of low-density settlement patterns. Layhe (1981) and Preucel (1987) both have applied the model to the ancient southwestern United States, although its predictions fit neither case; further applications appear in work by Hantman (1978), Effland (1979), and Dunham (1989).

Because the pattern of evenly dispersed, low-density settlement predicted by Hudson is entirely at odds with rural settlement in many areas of the world (P. Smith 1972),[2] an examination of the assumptions and causal arguments of the model are in order. The argument that in the second stage clustered settlement is produced by fissioning and dispersed settlement is produced by new immigrants warrants discussion. Hudson writes that the assumption "that successive generations show a limited spread away from birthplace . . . is such a widely used postulate in migration models in all sciences that it probably does not warrant further elaboration" (1969:370). This principle is indeed taken as axiomatic in many geographical studies (e.g., Morrill 1963b) and is a common, although not ubiquitous, feature of the settlement pattern (e.g., Hunter 1963). But why should it occur?

Bylund (1960) assumes daughter settlements stay near the "mother" settlement but leaves the reasons unstated. Siddle (1970) sees farmers as resisting fissioning in order to stay near their kin; when new settlements occur, it is the bonds of kinship that attract new settlements to old. Silberfein (1972:13) sees dispersion as being checked by the desire of "the entire family [to] remain together, close to shrines and ancestral graves." Bohland attributes kin settlement clusters in northern Georgia to "a strong sense of familism" (1970:20) and "a high frequency of interaction among family members" (1970:19), as does Jordan (1976) in Europe.

Common to these ideas is the assumption that new settlements will remain as close as possible to the parent settlement, but in fact the links between settlement patterns and social relationships are more complex. Social relationships such as household size and composition can promptly change in response to new settlement and labor situations (Netting 1965; Stone et al. 1984); under conditions of land scarcity, social relations may adjust to the economics of location rather than vice versa.

It is the third (competition) stage of the model that contains Hudson's major substantive contribution to settlement theory, and it is the implications of this stage that he tests against data from Iowa. However, Hudson's argument on the effects of land competition on settlement contains several assumptions that affect the applicability of the model. The assumption posing the greatest problem is that the nature of agricultural production does not change as population density rises. The relationship between agricultural intensity and settlement pattern is explored in some detail in the following chapter and so will not be pursued here. The second problem is that in the Hudson model, mature frontiers are characterized by populations stabilized at a low density, a condition quite unlike rural landscapes in much of the Third World. Grossman writes that "Bylund observed that technological improvements lead to greater efficiency of land use and, therefore, to *greater density* of settlement, whereas Hudson, drawing on the American experience of rapid urbanization in the past century, reached the conclusion that the final stages in settlement history lead to a *decrease* in the density of population" (1971:123, original emphasis).

For the third stage Hudson also assumes that small farms are absorbed by larger farms that tend to outcompete them. This is clearly not a general condition of farms competing for land, as the profusion of sustainable smallholder farms worldwide makes clear (Netting 1993). There are, however, particular conditions that favor economy of scale, including a reliance on capital-based inputs and the requirements of production regimes such as those favored in the Great Plains—cattle ranching and monocrop wheat.[3]

When we consider that the source of most farm inputs in subsistence economies is household labor, the assumption that large farms will try to absorb their smaller neighbors becomes questionable. For instance, Jackson (1972:258) notes that in tropical Africa, the usual case is that the "optimum size of holding is determined primarily by the size of the household"; Mortimore (1967) notes that in most of Nigeria, "the size of farm holdings is [a] function of size of labor force available at peak periods" (see also Norman 1969).

The third stage also assumes that farmers have somewhere to go, which may not be the case in many crowded situations. It ignores the possibility of changing the household economy in response to economic land pressure. Such change may be purely agricultural (intensification) or may involve a partial abandonment of the agricultural sector (e.g., taking jobs in town). Other alternatives include craft specialization and part-time farming with seasonal labor migration.

A further assumption of the Hudson model is that compact (nonfragmented) holdings will be maintained in the presence of rising population density, an expectation that is often invalid outside of the large-scale agricultural systems such as that used to test the theory (King and Burton 1982). Studies elsewhere have repeatedly documented the tendency toward farm fragmentation under rising population, in such diverse areas as Thailand (Hanks 1972), Bangladesh (Heston and Kumar 1983), Canada (DeLisle 1982), and West Africa (Udo 1961; P. Hill 1977:75; Hunter 1967). In fact, studies in the American Midwest, Canada, and England document increasing fragmentation even where farm size is on the increase (E. Smith 1975; Carlyle 1983; also see King and Burton 1982, Bentley 1987).

In sum, the model's assumptions make it inappropriate to most of the archaeological situations where it has been applied. I have critiqued the uses of the model here not because it is inherently flawed but because of its influence on archaeological thinking on the evolution of agrarian settlement; archaeologists have been insufficiently critical in their use of it. Moreover, the fact that so influential a theory of rural settlement wholly ignores agricultural change underscores the need for the investigation of the relationship between agricultural intensity and settlement location that is reported in succeeding chapters.

Summary

The models I have discussed seek to explain fundamental principles affecting rural settlement patterns, not to reach a detailed understanding of any particular case. This is the point of theoretical modeling as opposed to empirical studies. However, each of these models is predicated on assumptions about decision making that must be evaluated empirically.

The models of agricultural settlement discussed here deal with questions that have been left behind by mainstream geography (P. Hill 1977:62), where causal relationships in rural settlement of paleotechnic farmers have come to be neglected. Contemporary geographical approaches to settlement place a premium on the sophistication of mathematical models of profit maximization (e.g., Mulligan 1984) and point-pattern analysis (e.g., Haynes and Enders 1975; Thomas and Huggett 1980), often at the expense of consideration of causal mechanisms (King 1969:594). Even with the Hudson model, the main focus is on the mathematical properties of point patterns, although what archaeologists have seized on are the causes behind evolving settlement patterns.

Haggett (1965:24) attributes the geographical emphasis on mathematical modeling at the expense of theoretical understanding to geographers having been burned by their earlier proclivity toward cause-effect relationships in the days of environmental determinism. Thus Olsson's statement is as true today as it was when he first wrote it: Bylund and Hudson "have focused more on the spatial derivatives of the process [of rural settlement diffusion] than on the analysis of underlying economic and sociological factors . . . the existing theory of settlement diffusion is inconclusive and . . . the specific models which can be derived from it are essentially descriptive . . . explicit cause and effect relationships have not yet been established" (1975:85).

It is precisely these cause and effect relationships that are required to build a theory of agrarian settlement. At present we simply do not have this level of understanding, because our attention has been more concentrated on marketing and consumptive factors than productive factors (Schultz 1976:67). There is an important gap in our knowledge of how the productive activities of rural agricultural settlements affect the location, arrangement, size, and duration of those settlements. As a result, archaeologists have come to rely on models that hold agricultural production constant even as population density rises, a peculiar position given the state of agrarian ecology. Without a better understanding of the factors that actually drive agrarian settlements, it is impossible to adduce general models of agricultural settlement behavior. This, I would argue, is the cause of the failure of the Bylund, Siddle, and Hudson models to accurately predict and explain agrarian settlement.

The research reported in this book examines, in a carefully documented case of agricultural change, the relationships between agricultural production and settlement pattern. It contrasts with most of the studies discussed above in that it inductively seeks the factors that govern locational decisions as population density rises, rather than stating factors as a premise and deducing expectations for the resultant pattern. It is therefore on one hand a case study, an ethnoarchaeological view of the settlement process that I advocate for research on ancient settlement patterns. On the other hand, it is intended to isolate factors that actually drive agrarian settlement, as a step toward a general theory of agrarian settlement.

3

Agrarian Production and Settlement

My approach to agrarian settlement begins with relationships between population and agricultural change. Rather than starting from scratch and imagining what "rules" could have created a settlement pattern, I begin with what we know of food production systems and then link these agricultural considerations with settlement decisions. My experience with the Kofyar convinced me that any rules on where to settle are inextricable from rules on how to farm.

This chapter develops a production-oriented approach to agrarian settlement, treating farm settlements as integral components in a system of food production. The demands of production are not intended to explain all of the variability in agrarian settlement, but they do account for many of its fundamentals and also provide a foundation for understanding other aspects of agrarian settlement. In this chapter I explore fundamental linkages between agrarian ecology and settlement pattern and how this dynamic relationship is affected by population pressure.

Intensification Theory

The current field of agrarian ecology was shaped by Boserup's *Conditions of Agricultural Growth* (1965). In what she termed a "dynamic analysis embracing all types of primitive agriculture" (1965:13), Boserup reversed Malthus's view of the relationship between agrarian production and population. Where Malthus had seen the capacity for agricultural production as determining population ceilings, Boserup saw population increase as the independent variable determining agricultural change (1965:11). Traditional models of agricultural change were biased toward conditions prevailing when classical economists were writing, "when the almost empty lands of the Western Hemisphere were gradually taken under cultivation by European settlers, and it was therefore natural that they should stress the importance of the reserves of virgin land and make a sharp distinction between two different ways to raise agricultural output: the expansion of production at the so-called extensive margin, by the creation of new fields, and the expansion of production by more intensive cultivation of existing fields" (Boserup 1965:12). Models focusing on whether to establish new fields for permanent cultivation on virgin land were inappropriate for "many types of primitive agriculture [that] make no use of permanent fields, but shift cultivation from plot to plot" (Boserup 1965:12). Shifting cultivation with long fallows is a highly advantageous form of production (offering good returns on the farmers' effort and the amount of land in crops at any one time) although its production concentration is low. As population becomes more concentrated, so must production; forest fallow must be replaced by bush fallow, followed by short (grass) fallow, annual cultivation, and eventually multicropping. The key, as Turner and Doolittle (1978) pointed out later, is that production must not be reckoned on the basis of output per unit of land alone but output per unit of land and per unit of time combined. This is called production concentration.

Boserup did not assume that populations necessarily grow, but she focused on what happened when they did grow. As population density rises, so must production concentration. More people on the land means fallowed land must be returned to production more quickly, eventually to permanent cultivation. Agricultural intensification, then, may be initially defined as the process of increasing production concentration.

Raising production concentration changes the nature of farming, forcing farmers to contend with less fertile plots, covered with grass or bushes rather than forest, which requires expanded efforts at fertilizing, field preparation, weed control, and irrigation. These changes often induce agricultural innovation but generally increase marginal labor cost to the farmer as well: the higher the rural population density, the more hours the farmer must work for the same amount of produce. Labor efficiency here is the ratio of agricultural production output to labor input; it is because of this decreased labor efficiency that farmers rarely intensify agriculture without strong inducements.[1] As Brookfield summarized it,

> Strictly defined, intensification of production describes the addition of inputs up to the economic margin, and is logically linked to the concept of efficiency through consideration of marginal or average productivity obtained by such added inputs. In regard to land, or to any natural resource complex, intensification must be measured by input only of capital, labor and skills against constant land. The primary purpose of intensification is the substitution of these inputs for land so as to gain more production from a given area, use it more frequently and hence make possible a greater concentration of production. (1972:31)

What Brookfield captures is the importance of the relationship between inputs and land—as compared to writers who take intensification to be increasing production concentration regardless of inputs (Netting 1993:271) or define it as either increasing output or increasing inputs (Connah 1985:765). The Boserupian tendency, seen across time and place, is the decrease of labor efficiency in order to increase production concentration: rising rural population density necessitates increased production concentration, causing decreased input efficiency.

For each level of production concentration, there is a most efficient method of production; a line drawn through these optima is the intensification slope. In Boserup's formulation, rising population density is the cause of rising production concentration; this causes efficiency to decline, producing a negative intensification slope (fig. 3.1a). To Boserup, increased production concentration is necessitated by rising rural population density. For our purposes, however, it makes more sense to restate the driving force to be population pressure, or a rising ratio of food demand to the quantity and quality of productive land.

The fallow shortening that Boserup stresses is not the only mode of intensification. In addition to field labor–based intensification, with which her model is generally associated, it is helpful to distinguish capital-based and infrastructure-based intensification.

Capital-based intensification is characteristic of farming in industrialized societies. The amount of human labor required to produce a mouthful of food generally decreases, whereas the total direct and indirect energy costs of that mouthful climb to astonishing levels. The cost is paid not in labor but in cash (or credit); a contemporary single-family American farm may own a quarter-million dollars' worth of farm equipment and put substantial amounts of money each year into fuel, fertilizer, and pesticides. Indirect costs of this form of intensification range from the monies spent on both public and private agriculture-related research to the costs of military intervention to protect oil prices.

Although Boserup's original model of intensification stressed field labor intensification (see below), she later (1981:5) described just this sort of capital-based intensification.[2]

Infrastructure-based intensification is where the landscape is rebuilt to enhance, or remove constraints on, production. Growing seasons can be extended by cold-air drainage features (Riley and Freimuth 1979; Adams 1979), steep slopes can be terraced (Donkin 1979; Netting 1968), swamps can be drained or transformed into fertile ridges (Denevan and Turner 1974) or made into dikes (Padoch 1985), and dry lands can be irrigated via canals (Doolittle 1990). The work of constructing infrastructure, and much of the work of maintaining it, is often done between growing seasons. Such changes in the agricultural infrastructure are also referred to as "landesque capital" (Blaikie and Brookfield 1987; Kirch 1993), defined as land improvements designed to be used well beyond the present cropping cycle (Blaikie and Brookfield 1987:9).

In field labor intensification, marginal labor is invested in the land and crops with the expectation of raising output in the current cropping cycle. This is what is generally associated with Boserup's model, although she addresses infrastructure intensification as well (1981:55). Fallow shortening means more effort in field preparation, as the controlled forest fire is replaced by the clearing of bushes and scrub vegetation, by complete tilling (physical turning of the soil), and later by the plow, which requires the feeding and tending of draft animals. Weeding requires increasing amounts of labor, and field tactics ranging from transplanting to mulching to staking are adopted to boost productivity per unit of land. Irrigation may even be a matter of increased field labor rather than infrastructure development (an example being the pot irrigation practiced in formative Oaxaca; Flannery et al. 1967:158). Most of the work of field labor intensification occurs during the growing season, but the problems of labor bottlenecks lead farmers to a variety of creative scheduling solutions (Stone et al. 1990).

Boserup's model of population growth prompting intensification has fared well in the crucible of empirical testing.[3] What is harder to sustain is her argument that this process crosscuts areas of varying agricultural potential. Even Boserup's supporters have stressed that the emphasis on the common denominators in intensification overlooks crucial variations in environment (Sanders 1973:334; Brookfield 1968; see also Rubin 1973). The importance of environmental variation actually depends on one's level of generality. The fact that intensification, in its many forms, occurs in varied soils and climates is important to our understanding of fundamental relationships between agriculture and population—which is Boserup's level of concern. Yet at the same time there are sharp differences in whether and how farmers will intensify production depending on local ecological constraints; I will return to this important point.

Is Intensification Really Less Efficient?

The crux of Boserup's model—the drop in labor efficiency as production concentration rises—has been widely scrutinized. Studies over the last thirty years have supported it in broad outline, while also showing it to be something of a simplification. Consider, for example, capital-based intensification, which usually occurs where labor is relatively expensive. Although capital-intensive systems may offer extremely high output per unit of human labor, comparative studies that include nonlabor costs show that capital-intensive systems have some of the lowest efficiencies ever recorded (Leach 1976; Pimentel and Pimentel 1979; Bayliss-Smith 1982; for an overview see Netting 1993:123–145). Capital-intensive farming does not run counter to the Boserup model, but it involves factors well beyond her scope, because costs such as credit and gasoline are shaped not by local ecology but by regional, national, or even global political economy.

A more critical issue is whether labor-based and infrastructure-based intensification, where costs and rewards are mainly determined locally, can actually produce higher levels of efficiency rather than the lower levels in the Boserupian model. This was what Bronson argued (1972, 1975), and other important examples have come to light as well: wet-rice farming, which may be *more* efficient than swidden rice cultivation (Padoch 1985; Brookfield 1972; Turner et al. 1977); "intensive" raised fields believed to surpass the labor efficiency of swidden cultivation (Erickson 1993); arid areas where tree fallowing does not work (Smith and Young 1972; McGuire 1984); and Neolithic tree-fallow systems, which were inefficient because of the labor required by stone axe technology (Denevan 1992).

It is quite true that efficiency does not always drop as concentration rises, but measurement of the marginal costs of intensification is complicated because there may be qualitatively different kinds of costs in different types of agricultural systems. These are often missed. For instance, Padoch (1985) argues that Kalimantan farmers get higher marginal returns on labor in rice paddies than in swidden plots, but she does not consider the off-season costs of building and maintaining the pond-field infrastructure. In fact, shifting labor into the off season can be an important intensifying strategy (Cleave 1974; Stone et al. 1990) and is often a benefit of infrastructure-based intensification. By the same token, the plow allows the farmer to harness much more energy, boost production, and sharply increase returns on field labor; but it has substantial off-field costs, especially the keeping of draft animals (see Netting 1993:132–135; Pryor 1985; Pingali et al. 1987).

Agricultural change affects the type of costs as well as the amount of cost, and these changes must be considered as part of intensification. This does not mean that increasing production concentration necessarily lowers output efficiency. Local variation in the relationship between output efficiency and production concentration is one of the pivots of my approach in this study.

Other Causes of Intensification

Although there has been strong support for Boserup's model, there is clearly variation the model does not explain, and challenging Boserup has become a common way to enhance the theoretical significance of a case study. The pace of critiques of Boserup's model, or exposés of her assumptions, seems to be increasing rather than abating (Conelly 1992; Erickson 1993; Kalipeni 1994; Morrison 1996). To make sense of the enormous support for Boserup's model on one hand and the perennial criticisms on the other, we must recognize that the model is one of pattern and process at a high level of generality. Like all highly general models, it requires assumptions. To understand particular local situations, one must confront local factors other than demography and intensity; to understand particular phenomena that interact with agriculture, one must relax her assumptions.

In pointing to population growth, Boserup holds constant other causes of intensification; in pointing to intensification, she holds constant other effects of population growth. These assumptions on both sides of the equation require attention. My main interest here is in alternative responses to population, because many of these involve settlement patterns. Much has been written on factors beyond population that shape intensification. The most significant fall into the categories of risk aversion, social production, and market forces.

However common it may be for farmers to avoid wasting time and energy and to maximize profits from crop sales (Schultz 1964; Wharton 1969), a more pressing goal is avoiding food shortfalls (Schluter and Mount 1976; Kates et al. 1993) and building a cushion for bad years (Netting 1993:153). Kekchi Maya farmers raise production concentration by planting dry-season crops on levees, to mitigate risk rather than to compensate for disappearing fallow land (Wilk 1985). For beef producers like those described by Bennett (1969), conditions may favor expansion instead. At Teotihuacan, intensification appeared before being necessitated by population pressure, suggesting that agricultural intensification may arise to reduce risk in the face of environmental unpredictability (Sanders and Webster 1978); similar cases are made by Bronson (1972) and Nichols (1987).

Brookfield pointed out (1972:38) cases where production levels were "wildly uneconomic when measured by calorific returns, yet wholly reasonable when measured against social returns." Because food has social uses, social factors affect its production, and there are important social incentives for increased production that can directly affect agricultural intensity (Sahlins 1972; Brookfield 1984; Morrison 1994).

Market incentives can produce intensification in the absence of land shortage (Turner and Brush 1987; Netting et al. 1979). Market participation alters the farmer's basis for reckoning input efficiency, and when farmers are increasingly incorporated into a cash economy, they "make their production decisions in terms of pesos per hour, not kilograms per hour" (Eder 1991:246). Market participation can also expose the farmer to unpredictable price fluctuations, feeding back into the need to raise output to mitigate risk. The trade networks that archaeologists often see may have effects parallelling market opportunities; several studies have explored the effect of trade incentives on intensification (C. Smith 1975; Price 1977; Graves et al. 1982).

Modeling Alternatives: The Intensification Slope

Even if Boserup's model fits the broad outline of agricultural change, it is significant that sometimes farmers do not intensify; they allow yields to drop, or turn to craft production, or simply leave. The cross-cultural studies noted above may support Boserup's model, but they also indicate that something else is going on. Rather than simply chalking up these counterexamples as flaws in the Boserup model, we can make more systematic sense of variations in response to the need for greater production concentration by including information about local production. The model, as Netting (1993:276) points out, "becomes richer and more informative as it becomes more ecological."

Boserup's model is driven by farmers' responses to returns on agricultural inputs at the margin. Even within this paradigm, there should be considerable local variation in intensification. The fact that rising production concentration generally lowers input efficiency does not mean that each increment in every situation causes the same decrement in efficiency. Boosting production per hectare per year on a frontier with abundant land will not require the same additional tasks as squeezing the equivalent added production from an already intensified smallholding. This means the intensification slope is not the straight line shown in figure 3.1a.[4] Turner et al. (1977) found that population density correlated best with the log of agricultural intensity. They suggest this reflects a cross-cultural pattern in which "the rate of increase in agricultural intensity itself increases as population density rises. This result may reflect the influences of diminishing returns on the relationship. A point is reached at which a unit increase in population density will necessitate more than a unit rise in agricultural intensity if the production standards are to remain constant" (Turner et al. 1977:390). This would mean that the generalized intensification slope could be more realistically modeled as the curve in figure 3.1b. Although this is still a coarse and generalized model of the payoff scale to which farmers respond, it is an important refinement because it shows how even within Boserup's causal framework we should expect variation in farmers' responses to population pressure depending on where they are on the scale. Where the slope is flat it indicates that production can be boosted with little added input; where it is steep it shows that added inputs will have little effect on production.

The next refinement concerns inflections in the slope. Production concentration may be measured on a continuous scale, but input efficiency need not be continuous or incremental; it may have thresholds requiring qualitative changes in production. Let us look more closely at the slope to understand what is behind such slope changes.

The intensification slope is actually a composite. Each production regime has its own slope, and the composite slope comprises the portions of those slopes with the highest input efficiencies for each level of production concentration (fig. 3.1c).

For instance, slash-and-burn cultivation at its least intensive may involve the felling of few trees, basically to provide sunlight. With a slight

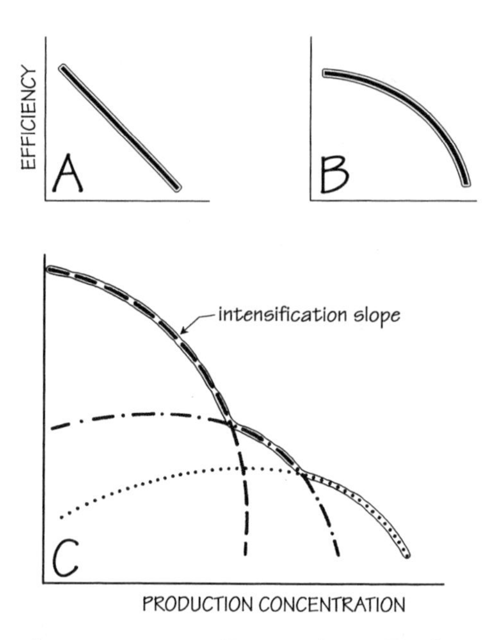

Figure 3.1. Intensification slopes. The figure shows three progressively more realistic models of the relationship between efficiency (marginal returns on agricultural inputs) and production concentration. The broken lines in part c represent three hypothetical production strategies that can be practiced in a given environment, such as forest fallow–bush fallow–grass fallow.

increase in labor, a firebreak can be felled, within which upright trees can be burned. More managed swidden systems may involve tying together a set of trees that can then pull down other trees that have been only notched, or individually felling all large trees in the plot. At the more intensive end of swiddening are tinder systems like those of the Maya, who keep Cohune palms for this purpose, or at the extreme, the *chitemene* system of southern Africa, in which tinder may be brought in from considerable distances to achieve a hot burn.

But most swidden systems reach a threshold beyond which higher marginal returns can be achieved by other farming methods. For instance, the returns on swiddening may be overtaken by returns on plowing. Plowing makes no sense for low production concentration because of its high start-up and overhead costs, but it can greatly increase production concentration when need be.

Thus, one production tactic might be pushed along the x-axis up to a point at which further production concentration exacts a much higher price on the y-axis. An example would be the point at which all the land surrounding a settlement has been farmed and it is necessary either to return to a short-fallow plot or to hike out to a more distant field. Another is the shortening of fallow to the point where enough ground-breaking is needed to warrant adoption of the plow, with the attendant costs of animal procurement and tending.

At this point there is an inflection in the intensification slope (fig. 3.1c), which may require not just a jump in inputs but other organizational changes required by new kinds of inputs. Downward inflections tend to be thresholds, the crossing of which may lead to major changes in agriculture (an example being the Indonesian woody fallows incapable of gradual intensification, which offer the farmer a discontinuous jump to intensive irrigated rice; Geertz 1963; Vasey 1979:270).[5] Guyer and Lambin (1993:839) point out that "for farmers as well as scholars, identification of these thresholds is critically important. Not only do maximum technical imagination and effort need to be concentrated 'at the cusp', but the qualitative change in production techniques also necessarily entails sociopolitical and cultural innovation: different labor allocation, different land access, different modes and temporal rhythms of sales and purchase." Inflections may affect decision making well before the production system reaches them. Farmers look ahead, and "the threat of overpopulation stimulates a variety of responses in order to put off a fall in output per head" (Grigg 1980:287).

Moreover, studies of population and intensification can be fooled by alternative responses to population pressure. Farmers may take action to counter the forces impelling them to increase production: if population pressure is rising, they may move, they may delay increasing their family size (Richards 1983:7), or they may eject others to forestall or prevent intensification. These responses are not what Boserup predicts, yet they maintain the population-intensity fit in apparent support of her model.

Localizing Slopes

Boserup's neglect of local agroecology was necessary at her high level of generality, but understanding farmers' decisions in particular cases—and the effect of these decisions on settlement pattern—requires building local agroecology into the theory. The intensification slope model is a useful conceptual tool in comparing modalities of environments and technologies at more specific levels.

Some of the most important differences between intensification slopes of different areas result from differences in the functioning of the fallows. Let us take the example of agroecology in the humid tropics, typically defined as receiving at least 1400 mm of rain per year with at least seven humid months (Troll 1966). Forest cultivation there is characterized by a steep slope reflecting very high efficiency for low-concentration swidden farming but precipitous drop-offs for more intensive methods.

A well-known property of tropical forests is that the preponderance of the nutrients are stored in biomass, whereas soils are relatively nutrient poor (Williams and Joseph 1973; Kowal and Kassam 1978). Unlike drier areas, where fields are burned mainly for clearing, burning in the tropical forest is essential in releasing nutrients for the farmer.

Plant and animal species—including weeds, fungi, insect pests, and parasites—flourish in the tropics, showing both high density and high diversity. Two aspects of low-concentration swidden/fallow cultivation are instrumental in controlling these biotic impediments. First is the burning that kills propagules, which explains why tropical swiddeners often take pains to ensure a much hotter fire than is strictly required for tree clearance. Second is the scattering of cultivation into discontiguous plots that shift through time, which is vital in preventing crop predators and diseases from becoming concentrated enough to decimate crops (Hecht and Cockburn 1989:ch. 5).

Tropical woody fallows characteristically rebuild humus and provide usable humic nitrogen at the same time as they accumulate usable minerals, suppress weeds, and possibly improve soil structure (Vasey 1979). In fact, the fallow itself is an economic resource, luring animals (Linares 1976) and providing resources (Hecht and Cockburn 1989:36–39).

As long as some woody fallow is maintained, production concentration can often be raised without high marginal input costs—for example, by bringing tinder into the field to raise the fire temperature, as in chitemene farming. But when production concentration is pushed beyond the point where woody fallows can be maintained, the drop-off in input efficiency is sharper than in dry tropics and temperate areas. Short fallowing encounters special problems in the wet tropics, where common grasses like Andropogoneae lower nitrogen levels and possibly humic content (as compared to temperate regions where grass fallows increase soil nitrogen [Vasey 1979:275]). Humid tropical farming, when population precludes woody fallowing and the best remaining option is permanent cultivation, exemplifies a downward inflection in the intensification slope.

Patterning in agroecology reflects these slopes; grass-fallow systems are more stable in temperate areas and relatively rare in humid tropics (Vasey 1979). We can work out slopes of increasing specificity, down to very local variation, reflecting how constraints on production differ along the x-axis for different landscapes. Soil sensitivity, or the extent to which soil can be favorably transformed by human intervention (Blaikie and Brookfield 1987), can vary sharply over short distances, depending on highly localized constraints on production. In the Namu Plains of Nigeria, where Kofyar farmers have moved, sandstone-derived sandy loams border shale-derived clay loams. On the sandy soils, problems of declining fertility and weed invasion can be met by incremental additions of field labor without a serious decline in input efficiency. In contrast, the clayey soils develop drainage problems that are scarcely ameliorated by incremental inputs, producing a more serious impediment to increased production concentration. The contrasting intensification slopes directly affect settlement patterns, as described in chapter 11.

Movement down the intensification slope may or may not involve technological change, depending on the impediments to production and the technology available. Gross (1975), for instance, classified Amazonian soils according to their "resistance" to swiddening (or the steepness of the intensification slope) and showed how this resistance affected the adoption of agricultural technologies.[6]

Local differences across the production continuum make it difficult to categorize land along a simple marginal/optimal scale. Two plots of land that offer equal returns to extensive farming may differ in their returns to intensive farming; plots that offer comparable returns to intensive farming may differ sharply in the minimum work required to bring them into production. As Brookfield (1972:42) points out, local agroecological conditions may offer "constraints that have the effect of providing 'threshold levels' of intensity below which no continued cultivation is possible."

The road to intensification followed by real farmers is not as smooth and linear as the Boserupian global highway; it has bumps and turns that vary with local conditions. To highlight the fundamental relationship between efficiency and concentration, Boserup averaged local patterns, which were neither gradual nor regular. The multiscale implementation of intensification profiles facilitates our modeling of responses to demographic and environmental change by particular groups of farmers in particular times and places in a way that is likely to add to rather than impeach our more general understandings.

The relationships represented by these curves do not in themselves determine agricultural change. Just as many factors promote increasing production concentration—such as population pressure, market demand, taxes, and the need to buffer annual fluctuations—there are many responses possible to the threat of lower efficiency, and variation in responses to downward inflections will be particularly instructive.

Settlement Shifting

So far in this chapter I have held settlement patterning constant; I have discussed farmers as making decisions about intensifying or not intensifying cultivation on a given plot. In real life, they will also consider cultivating different plots instead of (or in addition to) the given plot. We therefore must consider mutual influences between farming and settlement. Settlement pattern affects what other plots and levels of intensity are available; both intensification and opening cultivation on different plots affect settlement pattern. It is possible to deal with these mutual interactions in an integrated framework of agrarian settlement systems.

This topic is not easily separated from the problem of social organization of agricultural work. To now I have been principally concerned with the quantity of work required by the farm, but I will show that settlement patterns tend also to be sensitive to the kinds of work demands and the social means of meeting them. I begin with the fundamental issue of settlement shifting.

Faced with a descent in the intensification slope, the farmer's first response is often to move. Boserup's model applies to cases where this does not happen; she holds constant the distribution of human population except for population density.[7] Others have assumed the opposite, for example, offering flowcharts in which pressured farmers automatically remove themselves to a frontier when possible (Schiffer and McGuire 1982:269–272; Amanor 1994:66). But both assumptions obscure the vital link between agricultural and settlement decision making. Pressure or incentive to increase production concentration results in the choice of intensifying, moving, or ejecting population. Decisions to move all, some, or none of the population to another location are based on comparisons of the present location with alternative locations.

The keys to understanding the evolution of agrarian settlement systems are the criteria used in making these comparisons.

Shifting settlement can operate only so long as there is a relatively low ratio of population to productive farmland supporting the population. When yields on one plot decline, cultivation is shifted to other plots. If population growth is sufficiently high (or land regeneration sufficiently slow) to raise land pressure within the total cropland, there are a limited number of strategies for keeping production from falling below desired levels. For the intensification option, the cost (or the drop-off in the intensification slope) is often steep, and as Boserup writes (1970:101), "the first spontaneous reaction of tribal or peasant families to population growth within their community is to look for additional land to cultivate by the traditional methods." Shifting settlement systems are those in which farmers relocate settlements to avoid agricultural intensification.[8]

Shifting settlers are ethnographically underdocumented, accounting for only 1.2% of the agrarian societies in the *Ethnographic Atlas* (Murdock 1967). There are several reasons for this. In the first place, shifting settlement has been disappearing with the penetration of the market economy into subsistence agriculture; this can encourage intensification even in the absence of land pressure, and evacuation of agricultural surpluses, whether for market or other reasons, encourages settlement fixation. Cash from market participation commonly goes toward the structural improvement of residences and the creation of more permanent storage facilities, thus increasing the cost of abandoning the settlement. The normal time frame of ethnography is also poorly suited to cultural dynamics that may occur over a period of well over a decade.

But in the past, tropical areas with low land pressure were dominated by systems such as that described by Ruthenberg (1980:31): "In most systems with shifting cultivation, the continual movement of cropping results in a slow migration of the population. The cultivated plots move slowly away from the previous clearing and the vicinity of the hut. At the same time the cost of transporting the harvest increases. . . . Beyond a certain distance, it becomes advantageous to build a new hut near the field instead of carrying the harvest such a long way." Settlement shifting may be especially common in the tropics, but it is hardly restricted to the tropics. For instance, Netting (1989:223) notes that "our own Midwest was won by some rather shiftless shifting cultivators. Abe Lincoln's father went from one wilderness clearing to another, blaming bad luck but getting crops with a minimum of effort and moving when the hunting declined."

Shifting of settlement and shifting of plots around a settlement are responses to land pressure at differing scales. Land pressure on the highly localized scale of the individual farm develops very quickly, and this is what prompts the shifting of plots. Declining yields in a small *milpa* may be remedied by planting a nearby forested plot, and if all plots within a reasonable walking distance are played out but there are still forests beyond, the farm residence may be moved past its cultivated perimeter. It is when there is no land within an acceptable migration distance that something has to give. The question of what constitutes an "acceptable" migration distance is of course an empirical question, and an important one. Depending largely on the way they perceive (in my terms) the intensification slope of plots at various distances, farmers may choose among the basic options of settlement fixation, satellite settlement, and locational intensification.

Settlement Fixation

It is obvious why increasing permanency in settlement has concerned archaeologists over the years. Most attention has been paid to the various aspects of sedentism as a process linked to increased reliance on cultigens (e.g. Chapdelaine 1993). There is a comparable process within agrarian settlement that I call settlement fixation, defined as the strategy of retaining the settlement locus and altering agricultural production rather than shifting to a new location. This is the agrarian settlement strategy assumed in the Boserup model.

As settlement duration changes, so does the spatial organization of production. As shifting settlement gives way to fixed settlement, agricultural activities should become spatially segregated according to the intensity of labor input. It is widely assumed that "settlement location as well as sedentarization and settlement formation appear to be related to movement-minimizing behavior" (G. Johnson 1977:489) and that "farmers generally attempt to locate themselves . . . so as to reduce the distance between residence and field and minimize daily energy expenditures" (Sanders and Killion 1992:29–30); although this is not generally true of extensive farming, where the number of home-to-plot trips is low, it is increasingly true as agricultural intensity increases.

As land pressure increases, so must the radius of intensive cultivation, until the von Thünen principle becomes reversed: instead of land-use patterns adjusting to the distribution of population, the expanding margin of intensive cultivation and the time costs in accessing intensive plots force the population to adjust to the location of those plots.[9] This is the application of the proximity-access principle to agriculture: there is a premium on farmers' living near their fields that is commensurate with the access frequency required by intensification. Residences are "pulled" toward the plots by intensification.[10] Farm fragmentation can divide this pull so that there is little advantage in residing on any one plot; cultivation on consolidated holdings concentrates the pull and increases the attraction of the residence to the plot.[11]

It is important to make the causal relationships clear. Von Thünen treats agricultural intensity as a function of the distance from residence to plot, but distance from residence to plot is also a function of agricultural intensity. This is where we are left when we relax von Thünen's assumption of isolated settlement.[12]

In this approach to agrarian settlement, it is best to follow Chisholm in defining settlement dispersal functionally rather than as a formal characteristic of point patterns: dispersal is a measure of how close the population lives to the fields it cultivates (see Chisholm 1968:113; Demangeon 1927). Agglomeration is the opposite—residences are in close proximity, with little or no intervention of agricultural land.

In theory, a maximally dispersed settlement pattern would consist of individuals living on their farm plots. In practice, individual farmers almost never live on their own because the individual is an inadequate unit of production (although field houses may be at least temporarily staffed by a single individual). The most common form of highly dispersed settlement is the pattern in which small social units of production (normally called households) reside on or near their fields.

This pull also conforms to the tendency for land tenure to become established as cultivation intensifies, a tendency described by Boserup (1965:77–87) and others (Netting 1993; Udo 1963; Gleave and White 1969:280). As Smith puts it, "residence in close proximity to the cultivated land may not only be desirable because intensively cultivated plots require more care, but also because residential occupation is a way of asserting individual control over the land" (P. Smith 1972:415).

In other words, increasing population density should lead to each farmer's operating on less land and investing more labor, usually requiring more trips to the plot or plots. Increasing population in villages leads to expansion of the cultivation radius at the same time that it demands intensification. Farmers must spend more time on plots that tend to be farther away. This should encourage social units of production to be pulled away from the village, placing a premium on settlement dispersion.

The corollary is that when cultivation is not intensive and work inputs are low, movement efficiency exerts little pull toward dispersal. Dispersion is not necessarily efficient for agriculture, as is sometimes assumed (e.g., Steward 1955:167; de Montmollin 1989:299; Kohler 1992:622).

Dispersal need not entail great distances between residences, as the Hudson (1969) model predicts. There are several important factors acting to pull settlements toward each other, a phenomenon called settlement gravity. Dispersal and gravitation are affected by different phenomena and do not always vary inversely; agrarian settlements can be both dispersed and gravitative. Later I will look closely at agricultural collaboration as a cause of settlements' being attracted to each other.

In the framework developed in this chapter, rising land pressure forces an evaluation of the relative costs of intensification or mobility; land pressure results from more people, less land, less productive land, or a combination of these factors. Productivity of tropical agriculture generally declines rapidly, although the enhanced levels of solar energy and precipitation in the tropics offer considerable potential for intensification. Therefore land pressure is quickest to develop in the tropics, which, although favoring shifting settlement when regional population density is low, also support some of the most intensive systems of agriculture in the world.[13]

Land pressure also selects for fission or abandonment, whereas investment in local resources selects for residential stability and intensification. Here it is relevant that many of the same factors that contribute to the rapid development of land pressure in the tropics also contribute to tropical dwellings' being on the whole less permanent. In West Africa, the opportunity and material costs of establishing settlements were probably much lower than in the European nucleated villages with open field agriculture, because of "(a) a more congenial climate requiring less solid winter-proof structures, (b) the ready availability of relatively durable low-cost building materials, e.g., an abundance of woods . . . and lateritic mud for bricks and walls and (c) the fact that house building fits into the dry season, a slack period of the farm. . . . Thus we might expect towns and villages to grow, decline, locate and relocate with greater readiness in Africa than in Europe" (Richards 1978a:499). This also contributed to the favoring of abandonment over intensification.[14]

In the temperate zone, the generally higher initial soil fertility and gentler decline in intensification slope mean that, assuming no difference in population growth, land pressure should develop more slowly than in the tropics. The more favorable conditions for animal husbandry can lower the costs of intensification, providing further impetus for residential mobility or fission rather than abandonment.

Examples of relatively stable agrarian settlements in temperate areas with low population density are common. Early Neolithic villages in the Aegean (Renfrew 1972) and Japan (Watson 1977) are estimated to have had populations of a few hundred, and similar villages appear in the initial Neolithic in southern Scandinavia (Kristiansen 1982). A well-known example of agricultural colonization of the temperate forest is the Linearbandkeramic culture of central and southern Europe, with their small farmstead villages of longhouses, supported by a mixed economy. There is evidence for settlement continuity among the LBK even under conditions of low population density (Hamond 1981).

Studies of the dispersal of agrarian settlement through Germany in early historic times, based in part on place name analysis, support the model of settlement continuity on temperate farmscapes, with fission rather than abandonment occurring in response to land pressure. In northern Germany, for instance, village growth limited access to farmland, and small farmers, denied property rights in the Esch, "began at an early stage to leave the village group and reclaim new fields in the marsh and heath . . . and here they built their farms" (Dickinson 1949:247).[15] Population pressure in the Kilombero Valley of Tanzania, by contrast, has produced the classic trajectory of settlement fixation and spatial reorganization of agricultural activities (Baum 1968). Where land is most abundant, these rice cultivators follow a 45-year rotation cycle among three areas (fig. 3.2a). At each settlement there is a secondary cycle, as fields are regularly alternated before the settlement is shifted. Where land is less abundant, a 30-year rotation cycle is followed among two settlement locations (fig. 3.2b). Under further pressure, the family resides in a stationary settlement around which cultivation rotates (fig. 3.2c). The final stage is the intensive cultivation of the plots immediately adjacent to the settlement (fig. 3.2d).

When land pressure is artificially raised, the process is conflated, as in Cross River, Nigeria, when forest and game reserves so restricted the amount of land available that "shifting agriculture, with migrating settlements, was no longer possible, and it became more common to rotate cultivation among fields farmed from a permanent homestead"

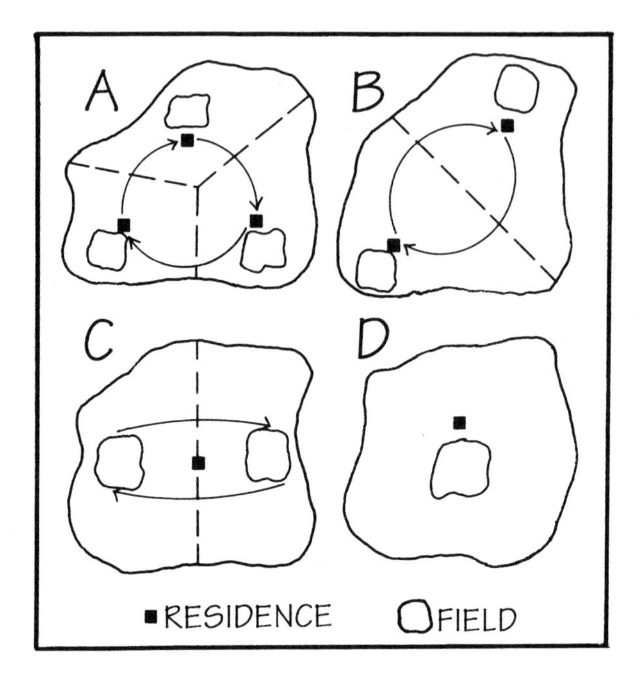

Figure 3.2. Schematic of the development of settlement fixation in Tanzania. Source: Baum 1968, with modifications.

(W.T.W. Morgan 1983:73). Although the maximum population density that can be supported by shifting cultivation varies with local ecology, Gleave and White (1969:275) offer the figure of 10/km² as an estimate for sub-Saharan Africa. However, shifting cultivation has also been reported where population density exceeds 50/km² (Turner and Brush 1987).

Many case studies in West Africa show the pattern described above, with villages surrounded by concentric land-use patterns in less densely populated landscapes, and dispersed settlement in denser areas. Grove (1961:125) states that the former settlement pattern is common where population densities range from 19/km² to 58/km²; W. B. Morgan (1969:316) concludes that the optimal development of land-use rings occurs in villages with populations of 2000–3000 cultivating an area 3–8 km in diameter.

Prothero's study of land use in Soba, northern Nigeria (1957), described a village with zones of concentric land use. Land within the village and within a belt roughly 1 km wide was manured and intensively cultivated. It was surrounded by a belt of land roughly 2 km wide and cultivated by rotational bush fallow, surrounded in turn by several kilometers of lightly cultivated fallow land. Schultz's (1976) later work in the same area concluded that increasing population density led to expansion of the intensively cultivated infield at the expense of extensive cultivation of the outfield.

Grove (1961) argues that continued population increase in such villages leads to farmers' seeking land outside of the initial zoned area, prompting a restructuring of the settlement pattern. Numerous cases from West Africa show the result of such restructuring to be a dispersed pattern. Among the Ibo, for instance, extensively farmed villages "disintegrated" into a pattern of dispersed, intensively farmed homesteads (Udo 1963, 1965; Jones 1945; Netting 1969). This pattern was demonstrated more thoroughly in Lagemann's (1977:28–29) comparison of Ibo areas of varying population densities. The low-population density village (100–200/km^2) is a compact settlement with farm plots extending radially from the town. Most of the area's oil palms (which reflect the permanency of a household's control over the land on which they are planted) are within the town, and a small patch of forest remains at the town's edge. The moderate-density village (350–500/km^2) is much less compact, with some of the area within apparently being used for crops. The farm plots around the town are smaller and usually have some oil palms. There is no remaining forest. The area of high population density (750–1000/km^2) has no village but instead dispersed compounds. Plots are smaller yet and invariably contain oil palms. In Lagemann's study, the intensity of cultivation corresponded neatly to dwellings' being located near the field.

Morgan (1957) describes the tendency in Iboland and Ibibioland (Nigeria) for areas with high population density to exhibit settlement dispersal, while sparser areas have small nucleated villages, a pattern he links to the labor demands of production. Gleave and White (1969) provide other examples of settlement dispersal in response to high population density.

There are examples from outside Africa as well. In China, "as cultivation intensifies it becomes increasingly rare for farmers to reside in agglomerated settlements away from their plots without a definite overriding factor" (Huang 1990); Huang (1990:64–65) compares northern Chinese villages, clustered on high ground to protect against floods, with settlement in the Chengdu Plain of Sichuan province, where there was no comparable threat and "villagers simply built their homes conveniently near their fields." In Guyana, the Akawaio farm the area surrounding villages of 20–60 people until yields decline, after which the group fissions into extended families living in separate "garden places" (Butt 1977). In Mexico, Rarámuri (Tarahumara) houses are located on or beside valley-bottom plots, which are farmed with the plow and manuring (Hard and Merrill 1992). Irish hamlets (*chachans*) gave way to isolated farmsteads when population growth caused the hamlets to swell; the dispersion process was pushed along by landlords as well (Johnson 1958).

Examination of contrary evidence shows how some supposed exceptions help prove the rule. Because, given the proximity-access principle, field labor intensification should promote dispersal, population pressure is often assumed to promote aggregation automatically. This linkage is sometimes attributed to density-dependent conflict (e.g., Orcutt et al. 1990), but often agricultural intensification is seen as the key. To Plog, for instance, the relationship between "intensified production and increased settlement size (nucleation) should be obvious. After all, the stimulus for intensification is population growth, and intensification makes further population growth possible. Moreover, intensification involves large cooperative projects. . . . Nucleated settlements are not essential for such projects, but they help, since it is much easier to mobilize people and get them to cooperate when they live close together" (Jolley and Plog 1984:439). Cordell and Plog (1979:417) also point to the organization of labor as a factor in the formation of large villages. Similarly, Vivian (1989:109) has stressed how the need for a large, coordinated work force to meet the demands of "horticultural intensification" promoted nucleation in Chaco Canyon.

It is true that the support of aggregated population normally requires intensive agriculture in the countryside. However, we should expect the primary producers to be pulled toward dispersion if farming is really field labor intensive. Where this pull is matched or overridden by a pull toward the town, we should expect the development of satellite settlements, as described in the next section. There are, moreover, quite satisfactory means of mobilizing large labor groups even where settlements are dispersed; this is facilitated by the formation of dispersed-settlement congeries, as discussed below.

It is also instructive to look at apparently contrary evidence from highland New Guinea, where a strong linkage between fixity of tenure and intensification or population density has long been recognized for groups such as the Chimbu (Brookfield and Brown 1963) and Mae Enga (Brookfield 1968). Yet Brown and Podolefsky's (1976) study of 17 highland groups shows correlations among population density, agricultural intensification, and individual land tenure but no correlation between agricultural intensity and settlement dispersal.

The most glaring exceptions to the pattern of intensification producing dispersion appear to be the Dugum Dani and the Gururumba. The Dugum Dani are said to have the highest population density in the sample (160/km^2) and nucleated settlement (hamlets). Yet the account used by Brown and Podolefsky clearly describes settlements that are dispersed by my definition: "The older and larger compounds are spread out along the edge of the valley with easy access to the forest behind, within sight of each other, and with their sweet potato gardens stretching out in front of them toward the frontier" (Heider 1970:49). The Gururumba settlement pattern does involve villages, and Newman provides a very clear description of Gururumba dispersed settlement: "some structures occur singly and some in clusters. . . . [T]he difference between houses occurring singly and in clusters is related to . . . [horticulture]. Most single houses are referred to as 'garden houses' or 'pig houses', dwellings erected within the fenced enclosure of a garden. Gardens are not always located near villages, and these houses save a person the trouble of walking back and forth between village and gardens during periods of intense gardening activity. . . . [P]eople may reside in either the village or the garden" (Newman 1965:18).[16] In fact, the New Guinea highlands offer a neat demonstration of the effects of intensification of settlement pattern that I have described.

Other parts of the world provide better examples of farmers living away from intensively cultivated fields, and these are worth looking at. Many parts of Europe have "agro-towns," agglomerated settlements where farmers may live more than 10 km from their fields (Blok 1969). One reason for this is that many farmers in agro-towns cultivate fragmented plots, which dilutes the pull of the proximity-access principle to any plot. A second reason is that the residential patterns of many European agro-town farmers, like those of the Gururumba, are quite dispersed. Whether or not the analyst or the farmer considers a house in an agro-town to be the farmer's "residence," intensive farmers often spend much of the time during the agricultural season near the plot, even if they are living in a secondary residence, which may be small or ephemeral. Even while calling the agro-town pattern "a paradoxical settlement which alienates the agriculturalist from the land he cultivates," Demangeon (1927) writes that "to reach the fields, the peasants must sometimes cover distances between twenty and thirty kilometers;

often they must pass the week far from their homes, only to return on Sunday" (Demangeon 1927:4, quoted by Blok 1969:122; my emphasis). In other words, for six days out of seven, the farmer is living on the agricultural plot and is by definition dispersed! This is a prime example of satellite settlement, which is an important settlement response to intensification.

Satellite Settlement Systems

I have argued that a corollary to von Thünen's theory of land use is that an effect of agricultural intensification is the pull of individual settlements toward the associated farmlands. This is not to say that all intensive farming systems have dispersed settlement; they do not. Rather, agricultural intensification places the premium on field/residence propinquity, because more trips need to be made to the field and because the increasing investment into less land promotes the establishment of individualized tenure.

This pull toward dispersion may be overridden. Military threat can force settlement agglomeration (e.g., Udo 1965; Rowlands 1972), just as it can promote settlement fragmentation (Beckerman 1987:73). Reliance on fields in diverse edaphic settings (field fragmentation) can also eliminate the advantages of living on or close to any one plot (Galt 1979; Bentley 1987; King and Burton 1982:482). Dispersion may also be overridden by attraction to central places. Note that the crux of central place theory is that the distance between population and central places adjusts to the rate at which the population accesses those central functions. Equilibrium is approached not only by developing central functions in areas closer to the client population, but by attracting the client population to where the functions are offered.

The Yoruba of southwestern Nigeria provide a classic case of satellite settlement. Despite moderately high population densities and some intensive agriculture, the Yoruba lived mostly in nucleated towns with large populations (Mabogunje 1962; Ojo 1966). Although defensive considerations provided the original impetus for nucleation, attraction to central functions now plays a larger role. The Yoruba maintain city residences only at a considerable cost, as shown by studies of time spent in travel to their farms (Ojo 1973). However, although conventionally considered to have a nucleated settlement form (Murdock 1967), they are increasingly residing on their farms for most of the agricultural season (Ojo 1973).

The factors favoring satellite settlement systems are shown most clearly where nucleation is forced onto a dispersed, intensively farming population. In 16th-century Peru, dispersed farmers were forcibly removed to *reducciones,* from where they commuted to farm intensive plots: "Gradually the peasants built houses near their plots in order to guard the crops. . . . The satellite settlement differed in appearance from the original village. The 186 dwellings were . . . spaced between 20 and 100 meters apart and were located next to the parcels that were individually owned and farmed. The degree of dispersion was largely determined by the maximum farm size and the need to protect investments" (Gade and Escobar 1982:441).

Whatever the factors favoring agglomeration of intensive farmers (see Glassow 1977), the result is a spatial contradiction—a pull toward the farm for agroecological reasons and toward the nucleated settlement for social, economic, military, or other reasons (Bunge 1962; Richards 1978b). This spatial contradiction is the agrarian analog of Binford's (1980) "spatio-temporal incongruity" among hunter-gatherers, and the solution of satellite settlements on farm plots is analogous to the hunter-gatherer solution of logistical organization. But where Binford treats logistic (special-purpose) sites as qualitatively different from home bases, agricultural satellites and "home bases" are on a continuum, varying in intensity of occupation. At one end of the spectrum of "use intensity" are the Iban, who have been used as an analog for the early Maya (Voorhies 1982:67). When the Iban are residing in *dampa* satellite settlements, the dampa is their full-time residence, and their longhouse is essentially abandoned for many years at a time (Freeman 1955). The Akawaio of Guyana, who live mostly in "garden-place" satellites, visit their more enduring villages occasionally. Most field house satellites in the prehistoric American Southwest were probably occupied on a seasonal basis (Kohler 1992). The Kofyar "guard house" on outfield plots, which serves as little more than a refuge from rain, lies at the least intensively used end of the spectrum.

But settlements can move along the spectrum; satellites may evolve into homes. This may result from the variables of distance to plot and agricultural intensity as described above: as more labor is invested in the satellite farm, there is increasing utility in the workers' spending more time close to the plot. With this also come greater demands for nonagricultural labor, as in the Yucatán, where "men would set up *champa*s, or temporary shelters, at distant milpas where they remained the entire week and returned only for the obligatory Sunday mass. At some point, possibly because they got tired of cooking their own food . . . and washing their own clothes, they would move the family and household out to the milpa" (Farriss 1978:212; see also Sanders 1967).

Locational Intensification

Shifting settlement systems can avoid intensification only so long as there are fertile areas into which settlement can move. As these areas run out, farmers may intensify as Boserup describes, which often pulls the residence nearer to the plot either permanently (dispersal) or intermittently (satellite settlement). But extensive cultivation can sometimes be maintained only by heading for poorer areas where farming could be successful only with added inputs.

How this agrarian settlement phenomenon relates to Boserup's version of intensification is ambiguous. For instance, the soil and water control features around the Mogollon site of Chavez Pass were convincing evidence of ancient agricultural intensification to Upham (1985). However, Reid (1985) protested that the features did not indicate intensification, because they were a necessary and predictable response to conditions of the site: one could not farm there without such features. By the same token, Feinman et al. (1984) describe increased use of marginal piedmont lands in Oaxaca as evidence of intensification. Is it "intensification" when farmers turn to land where cultivation is impractical without extra effort?

The question has been raised by other writers (e.g., Padoch 1985: 273; Turner et al. 1977). As Grigg (1979:77) put it: "Nor is irrigation or the length of fallowing solely a response to population density. In many parts of the world crop production is impossible without some form of irrigation, and historically would have occurred at the beginnings of agriculture rather than as a result of a long process of increasing population pressure, reduced fallow, and finally, the adoption of irrigation to allow multiple cropping."

The issue can be clarified in terms of localized intensification slopes. Localities differ not only in the shape and slope of their intensification profiles, but in the left end of the profile, which is the input efficiency for low production concentration. Steep hillsides that cannot be cultivated without terraces, drylands where crops need irrigation, and swamps without plantable surfaces have intensification profiles that start relatively low. Movement into such areas is called locational intensification.

The findings of Sanders et al. (1979) in the Basin of Mexico exemplify locational intensification. Here, "initial colonization would occur in the humid south and would involve extensive systems. Under population pressure two processes would occur simultaneously: increasingly intensive cultivation of the low-risk lands of the south and population expansion into more arid northern areas where more intensive land-use techniques (i.e., terracing, irrigation, and drainage cultivation) would be used even early in the colonization process" (Sanders and Nichols 1988:43). Doolittle gives a concise statement of how locational intensification is mediated by labor efficiency:

> Ideally, . . . agriculturalists should utilize extensive, labor-efficient techniques on optimal lands to satisfy production demands in nonstressful situations. As population pressure increases, agriculture should be expanded throughout optimal lands without technological change. When demands may no longer be satisfied within this option, agriculture should be intensified to increase output on land already under cultivation, on previously unused, less optimal land, or by some combination of these methods. The specific response should be the most labor-efficient in the context of local conditions. (Doolittle 1980:329; see also Hansen 1979 and Feinman et al. 1984)

The "ideal" described here is the ecological baseline with which others interact; the point is not that agrarian settlement is exclusively responsive to labor efficiency. I have outlined ways in which residential mobility can mitigate the reduction in efficiency brought about by intensification. Locating settlements farther from domestic water in order to be near productive land is comparable to locational intensification: Boserupian pressure is causing a settlement change that creates an increased workload. This process is examined empirically in chapter 11.

Agriculture, Labor Demands, and Social Groups

The foregoing has emphasized quantity and efficiency of work inputs. But in systems where these inputs are generally not purchased, they must be mobilized socially, a process affecting both overall social organization and settlement pattern. The social organization of labor involves not only the primary units of production like households, but collaboration among these units. Labor pooling is often a basic strategy, not only in episodic events like erecting buildings (e.g., Boatright 1941), but in the ongoing demands of agricultural production, even where settlement is dispersed.

Most paleotechnic agriculture, characterized by its reliance on human and animal energy (Wolf 1966:20; Turner and Brush 1987), has long been a partly collaborative affair, with farmers regularly assisting in the production of others' crops. Agricultural collaboration, which turns out to be important to settlement on several levels, has been misread repeatedly. Marx was unwilling to recognize collaborative labor unless it was based on shared ownership of the means of production (Donham 1990:13); both Chayanov's (1966) and Sahlins's (1972) models saw the farm household as producing independently. Writers have repeatedly reported cooperative labor to be disappearing, a casualty of the rise of "individualism," dissatisfaction with the quality of the work, technology, monetization, and the emergence of a rural proletariat. Yet, although there are certainly cases where cooperative labor has dwindled, in many others the imputed drop-off may have less to do with the social organization of labor than with nostalgia for a past in which cooperation has been exaggerated (Moore 1975). My survey of the literature finds it to be nearly ubiquitous in paleotechnic systems and surprisingly common in neotechnic systems.

Much of the logic of cooperative labor was outlined by Erasmus (1956). His distinction between exchange labor (groups normally under 10, working on a purely reciprocal basis) and festive labor (much larger groups, recompensed with drink and often food) has held up well. Labor groups can complete tasks when time is tight or when several operations need be done simultaneously or when unexpected labor crunches need to be met. Local group labor may promote a faster pace of work (fig. 3.3) and is invaluable for agricultural tasks requiring the application of many hands at the same time—simultaneous labor demands, as opposed to linear labor demands, which can be met by consecutive inputs of labor by a small group working for a longer time (Wilk and Netting 1984).[17] Festive labor also has an important social aspect, allowing the farmer to convert some of the subsistence economy's perishable capital into prestige.

Agricultural labor groups occur worldwide, from Brazil to Bulgaria, from the Trobriands to China (Erasmus 1956). They occur in paleotechnic and neotechnic systems alike, including contemporary England (Appleton and Symes 1986) and Canadian Hutterian communities (Bennett 1967). A study of farming in rural Michigan in the 1940s found that "men band themselves together in co-operative activity around planting, harvesting, threshing, and storing of farm products" (Kimball 1950:38). Backwoods pioneers on the eastern seaboard of the United States had formal work associations (Jordan and Kaups

Figure 3.3. A festive labor party (mar muos) in Ungwa Kofyar, complete with a drummer to encourage the workers.

1989:47), and settlers in Misiones province in Argentina formed "societies" for farm tasks like tree-stump removal (Eidt 1971:135).

Although the pervasiveness of labor pooling may be recognized, many of its ramifications have not been appreciated. The failure to consider these social entailments of agricultural labor demands has been one of the most common causes of problems in agricultural development projects (cf. Maos 1984). Agricultural collaboration also turns out to be a key factor in understanding agrarian settlement. To understand why, consider the costs of collaborative labor. First there are the time and energy costs of travel. The proximity-access principle applies just as much to accessing neighbors' plots as it does to accessing one's own plot. Agricultural collaboration turns out to be a prime source of Kofyar settlement gravitation; both compound spacing and farm shape adjust to this interfarm movement.

Agricultural collaboration also costs some control over one's own schedule, and I know of no agricultural system, prehistoric or modern, in which scheduling is not critical. Because farmers trade in labor, and because schedule control is generally negotiated locally and without legal contract (cf. Robertson 1987), participants must enter into arrangements sharing notions of farming, labor, reciprocity, and censure. "Participation in mutual aid societies," as Williams writes (1977:77), "depends on the ability to manipulate relevant symbols and participation in noneconomic social and cultural activities." This causes ethnicity to take on considerable importance in settlement decisions. Chapter 10 examines this close and complex relationship between social and spatial propinquity on the Kofyar frontier.

Discussion

I have explored an approach to agrarian settlement that begins with a theory of agricultural intensification, then reshapes that theory to allow it to better adapt to local ecological variation, and then relaxes its assumptions about settlement to allow examination of agrarian settlement systems. The phenomena discussed in this chapter certainly do not "determine" all agrarian settlement patterns, but they do produce constraints and tendencies for subsistence and peasant farmers cross-culturally. To make sense of variability in agrarian settlement patterns we have to look at how these relationships between agricultural production and settlement are played out in particular historic contexts.

The point is not to insist that agrarian settlements conform to ecological optimization but to build an understanding of the agroecological constraints and pressures that must be overridden. Where such constraints and pressures are mild, it may take little to override them. Rutter (1971), for instance, may have something in claiming that the Ghanian Ashanti's nucleated, deforested villages of courtyard houses result from a psychological need to control a threatening environment. If the ecological cost of agglomeration is slight enough, I see no reason why the "feeling of dread" that he describes (or more likely the benefits of a village's sociality) could not override it. The opposite happened in the American South following emancipation; freed blacks left what Aiken (1985:391) describes as the economic advantages of agglomerated quarters to reside on individual plots, in what boiled down to a "spatial expression of freedom."

From the general I will now move to the particular, looking at how the relationships described in this chapter were played out as Kofyar farmers settled the agricultural frontier south of their homeland.

4

The Kofyar Homeland

The subject of this book is what has happened on an agricultural frontier after a population began to settle it in the middle of this century. In this chapter I describe what was happening in the area those settlers came *from*—the "homeland." I concentrate especially on the same topics that I analyze on the frontier, namely, agriculture, the social organization of production, and settlement pattern.

Geologic Background

The Jos Plateau owes its existence to events surrounding the breakup of the ancient supercontinent of Gondwanaland. When the South American landmass sheared off the African Shield below present-day West Africa, a zone of tension followed the western margin north into the remaining landmass, leaving a crustal arch with concentrated magmatic activity. As magma breached the weakened crust during the Jurassic, granite intrusives formed in central Nigeria, in the Matsena area of Bornu, and in Niger around Zinder and the Air Plateau (MacLeod et al. 1971:48; W.T.W. Morgan 1983:6). These durable Younger Granites protected the surrounding landscape from 160 million years of erosion, leaving a ragged series of uplands. The Jos Plateau is the largest, covering 50,000 km^2 near the center of Nigeria (fig. 4.1). Most of the plateau surface is 1000–1200 m in elevation, but elevations surpass 1500 m in places. The Younger Granites are still visible as isolated hill masses rising above the more ancient platform, which is composed of Precambrian and Cambrian crystalline rocks known as the Basement Complex.

The hills of the plateau offered protection to groups such as the Kof-

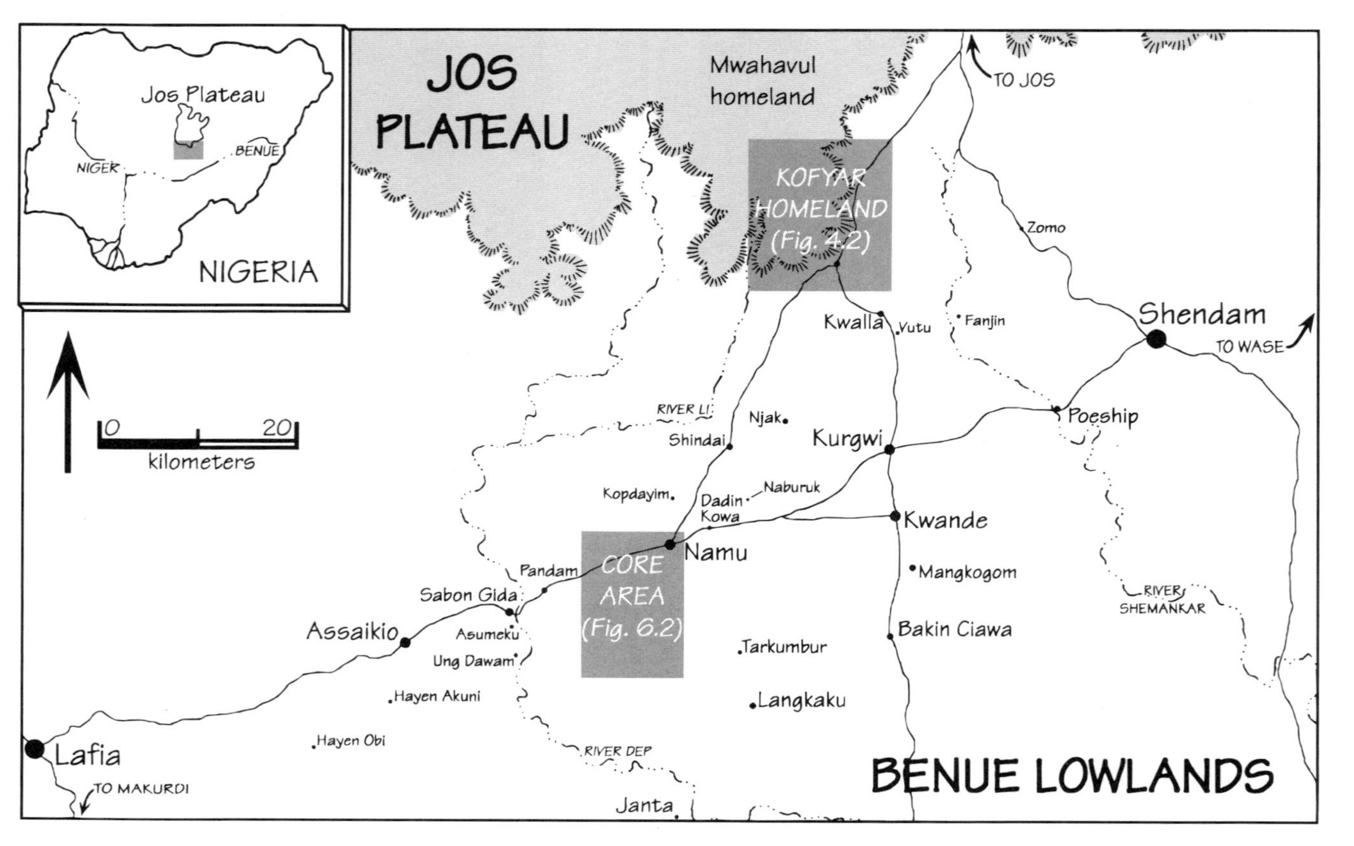

Figure 4.1. The southern Jos Plateau and part of the Benue Lowlands.

yar, who established a homeland in the southeastern corner of the plateau and on the plains immediately surrounding the base of the plateau. The plateau landscape here is rugged, with dramatic hills and steep-sided gorges. Rivers are deeply incised along joints in the Basement Complex, running predominantly northwest to southeast, with the principal stress directions of the plateau (Ajakaiye and Scheidegger 1989:727). The edge of the plateau is an irregular escarpment dropping several thousand feet.

Mid Tertiary volcanic eruptions have left volcanic features throughout the area, including two large cinder cones and the Moelaar, the volcano overlooking the village of Bong. There is a large flow of vesicular basalt boulders to the east of the plateau.

Origins of Homeland Settlement

There is no archaeological information available for the Kofyar homeland, and so for questions of when and why the Kofyar population came to be concentrated there we must turn to historical accounts. Because the Nigerian Middle Belt has long been seen as a cultural backwater between the Hausa city-states of the north and the southern civilizations such as Benin, these accounts are thin.

Population movements into the hills of the Jos Plateau were basically a defensive reaction by relatively small groups to military threats on lowland areas. The northern end of the Middle Belt was home to many of these groups, which lacked military strength to resist the slave raiding of the Zaria Hausa state and the Fulani jihad (Gleave 1965:127; see also Hogben 1930; Meek 1931; Buchanan and Pugh 1955:80). The hills were "beyond the reach of the Fulani who, with Wase as their base, carried out occasional raids as far west as Shendam, and it was more or less free from the attention of the Lafia raiders who harassed Namu, its southern neighbors" (Findlay 1945:139). The precipitous climb up the escarpment also provided a barrier to cavalry.

The long history of military danger on the Muri Plains below suggests that the Kofyar have resided in the Jos Plateau homeland for several centuries. Temple opined that "they migrated to their present location, circ. 1830–40, to escape Filane [Fulani] pressure" (1919:277). Where they were before that is open to speculation; Findlay (1945) assumed they came from higher on the plateau.[1]

Kofyar settlements also occurred below the hills, and by the turn of the century, a substantial population was living on the plain surrounding the escarpment (fig. 4.2) in areas called Merniang, Doemak, and Kwalla.[2] It is likely that the plains settlements were spillover from the hills. The local assertion that settlement spread from the hills to these plains areas is consistent with the Kofyar/Njak/Doka origin myth, which traces humanity to the hill village of Kofyar or Kofyar-Paya, from where Dafier and his sister Nade began to populate the world. Language also hints at the priority of hill settlements: the Kofyar word *koepang*, which equates to the English *home* and *house*, literally means "of the rock" or "of the mountain."

The colonial officer Fremantle believed the Doemak and Kwalla to have moved onto the plains because of pressure in the hills, offering the date of 1835 for the establishment of Kwalla (1922:46–47). They were certainly established on the plains by the turn of the century; British troops under Captain Ruxton subdued Kwalla in 1909 "after a spirited fight" (Fremantle 1922:47; Fitzpatrick 1910).

If the protection of the hills was important, it is curious that settlements were as far out as Kwalla—8 km from the escarpment. Part of the answer is that plains farmers could have retreated to the hills for protection, and at least some plains communities maintained long-term links with hill communities. Many of these "hill-foot" Kofyar lived near flows of vesicular basalt that would been largely unnavigable by horse-borne raiders.[3]

Population and Production

Land pressure was generally high in the Kofyar homeland, but there were differences between hill and hill-foot (plains) communities (fig. 4.2). In 1945 the density on the Doemak plains was estimated at 1200/mi^2, or 463/km^2 (Findlay 1945). This is consistent with my estimate of 488/km^2 for the population of Merniang in 1957. Population density in the hills was estimated at 96/km^2 in 1945. Using Netting's censuses from 1961–62 and 1:40,000 aerial photos from 1963, I calculated population densities ranging from 37/km^2 to 100/km^2 in hill communities, taking into account outlying agricultural areas (Stone et al. 1984). Yet much of the hill landscape was in steep slopes and covered with thin soils, so that land pressure was higher than simple density figures suggest.

Netting (1968) has described Kofyar homeland farming in detail; I will only summarize it here. In both the hills and plains, cultivation was highly intensive. In most villages, the infield surrounding each com-

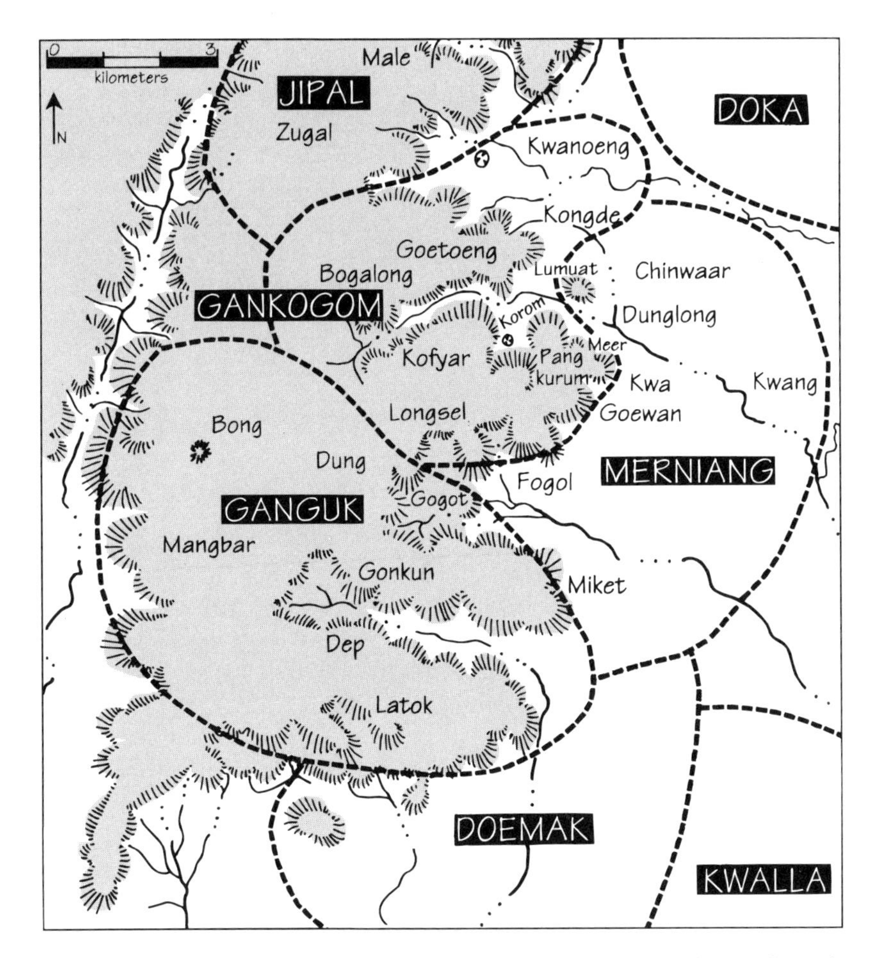

Figure 4.2. The central Kofyar homeland. Sargwat (warfare alliance) boundaries are shown by the heavy broken line.

pound was put in intercropped sorghum, early millet, and cowpeas each year; villages at the highest elevations relied more on maize and late millet. The infields were fertilized annually with compost from stall-fed goats (which was hoed into waffle gardens to control infiltration and check erosion) and were carefully weeded. Seedlings were often cultivated in nurseries and transplanted by hand. A wide roster of secondary crops—including maize, groundnuts (peanuts), bambara nuts, cocoyam, sweet potato, dwarf millet, sesame, and various cucurbits—was grown on terraces and outfields. The Kofyar also cultivated palm and canarium trees for their oil, which was both consumed and sold in local markets.

Although the infield was responsible for the lion's share of agricultural produce, the Kofyar also cultivated three categories of outfields. Plots cultivated on village perimeters or interstices were called *mar lang*, and plots that were more distant (but usually within a half-hour's walk) were called *mar goon*. Outfield plots were cultivated extensively, usually with swidden methods and fallowing, but light fertilizers such as ash were sometimes applied. The last category of outfield was migrant farms (still usually within a day's walk) called *mar wang*, discussed in chapter 5.

Many hillsides were terraced. Most terrace plots were outfields, but some compounds were situated on slopes so steep that the infield (*mar koepang*) was entirely on terraces. Cultivation tactics on such infrastructure can involve low labor investment and fallowing, which are hallmarks of extensive farming. Yet in most cases where agricultural infrastructure has been built, the cultivation regime is indeed intensive in that labor is being substituted for land (see chapter 3). Labor inputs on terrace plots were relatively low; they were only lightly fertilized, and fallowing was responsible for restoring fertility. Yet the time cost to build and maintain the terraces was quite high. In fact, when new lands opened up to the Kofyar, it was the terrace plots that most farmers abandoned, and on the steeper hills the unmaintained terraces began to disintegrate quickly.

The great majority of agricultural labor was mobilized within the household for work on the infield. Netting characterizes the labor demands of this agricultural regime as "small-scale but continuous" (Netting 1965) and manageable by the relatively small labor pool of the homeland household. Indeed, additional workers in the household offered little marginal utility because production was limited more by land than by labor. The characteristically small Kofyar household is therefore well adapted to the labor demands of production, and the swollen households that were kept from fissioning by land shortages in plains villages were disadvantaged (Stone et al. 1984).[4]

Suprahousehold labor arrangements also played an important role. The two principal types were *wuk* (reciprocal labor groups) and *mar muos* (festive communal labor parties). Mar muos were large, festive work parties for which contributors were compensated with millet beer and often food.[5] They were uncommon on the crowded plains but frequent in hill villages. Netting recorded an average size of 38 workers in

Bong village. Wuk groups, usually comprising 10–20 workers, would meet sporadically to work on one another's fields. The work was compensated only with reciprocity, which was important at times in the agricultural cycle when millet for beer brewing was in short supply (Netting 1968:136).

Sociosettlement Units and Production

I have held that agriculture must be seen not only as an ecological act, but as a social process. On the frontier, settlement pattern is tied closely to the social organization of production, which is in turn embedded in patterns of group affiliation. The three phenomena are nearly impossible to isolate; they are, in a sense, inherent in one another. Before we can ask how social affiliation affects locational decisions on the frontier, we have to deal with the social affiliations that migrants brought from the homeland.

In doing so we are pulled inevitably to the concept of ethnicity, even if it has become too laden and loaded an idea to work well in describing the diverse patterns of behavior, affiliation, and identification in many other contexts. The designation of an "ethnic group" may be less unambiguous when a set of people are culturally homogeneous, believe in a set of identity symbols, are organized politically, and share a self-concept. Yet often this is not the case, as with the many American Indians who shared linguistic and cultural traits but who developed political integration and a shared concept of identity only through interaction with the United States government (Cornell 1988). The criteria of shared self-concept and self-identification, often privileged in assessing ethnic status, may be absent within social groupings that are vital in the shaping of behavior. By the same token, people may self-identify as belonging to social groupings invented by colonial administrators based on boundaries that do not fit patterns in behavior.

Understanding settlement locational decisions demands an examination of how social affiliations affect work organization; in this I will make use of the concept of social propinquity, a property manifested in shared values and expectations that are conducive to agricultural collaboration. The concept is not problem free, but it avoids some of the problems of "ethnicity" and can guide us through a welter of social taxa to an approximation of patterns of social affiliation that are meaningful to the organization of production.

In the Kofyar homeland, social affiliation is expressed through the id-

Table 4.1 Household and Compound Sizes in the Kofyar Homeland

	Cases with Huts				Entire 1961 Sample	
	Household Size	Huts	*n*	Huts/ Person	Household Size	*n*
Hills	4.6	9.3	298	2.0	4.8	410
Plains	6.4	7.7	179	1.2	5.6	302
Overall	5.3	8.7	477	1.6	5.1	712

Source: Based on data collected by R. Netting in 1960–1962.

iom of geography. The Kofyar language admits no term like *tribe* or *ethnic group*. One asks about social affiliation by asking *Ga gurum pene*? which is not a "who" question but a "where" question—literally, "What place are your people from?" The answer is invariably a location, the specificity of which varies with the circumstances. Like the Diola described by Linares (1983:130), the Kofyar recognize social relations in spatial terms, and Linares's concept of the sociosettlement pattern is very apt here. I first define the various sociosettlement taxa of farmstead, neighborhood, village, alliance, and tribe, and the relations of production within them; then I look at social propinquity as reflected in endogamy patterns.

The Farmstead

The elementary unit in the homeland settlement system is the farmstead (*mar*), consisting of a residential compound (*koepang*) and an infield area (*futung* or *mar koepang*) that usually surrounds it. Most farmsteads have their own name and history and relatively stable borders.

Hill compounds are arranged irregularly, often perched on high points, whereas plains compounds are more evenly arranged. Hill farmsteads are often separated by unfarmable hills and stream valleys; plains farmsteads are contiguous. Netting (1965) calculated an average farmstead size of 1.55 acres (0.63 ha), but it is difficult to compare sizes of hill versus plains farms.[6]

Each farmstead is home to a household, most commonly a polygynous, nonextended, co-residential family (Netting 1965, 1968; Stone et al. 1984). The household's residential compound (koepang) in 1961 had an average of 8.7 huts[7] and housed an average of 5.1 individuals, but there was a difference between the hills and plains (table 4.1).

Households were considerably larger on the plains, but compounds were slightly smaller. As a result, plains households had a much lower ratio of huts to people. The reasons have to with the relationships among houses, people, and land.

In the first place, compound size in the homeland is much more sensitive to household growth than to household shrinkage. For instance, when a new wife enters the household, it is almost always possible to erect a new hut for her, but the hut will rarely be torn down when she leaves. A hut may be maintained as an extra bedroom, kitchen, or storage area, requiring only that the roof be rethatched periodically. Therefore, compound size correlates with household size only when households are at the peak expansion of their developmental cycle (Stone 1983). Small households might have anywhere between 0 and 6 extra rooms. Plains households spend much more of their developmental cycle in the stage of greatest expansion. They have an "impacted" developmental cycle, due to shortage of the land needed for household fission (Stone et al. 1984).

Before the days of migrant farming, there was a near-perfect correspondence between the settlement unit of the compound and the social unit of the household. The exceptions were few. Sometimes a household fragment (such as a wife and an adolescent son) would move into a recently inherited farm and begin to work it while continuing to participate in other ways in the household's affairs. Also, women who moved in with a *wumulak* (medical/religious practitioner) to give birth often remained there for several years, acting as partial members of both the wumulak's and their own households. Otherwise, the co-resident group on the farmstead was the social unit of production, and the great majority of farm labor was accounted for by household members working on the farmstead or its outlying fields.

This changed when households began sending some of their ranks to migrant farms to the south. In most cases, the migrant farmers were still very much members of the homeland household.

The Neighborhood

Groups of farmsteads form neighborhoods the Kofyar call *toenglu*. *Lu* is a house or home, and *toeng* is a tree or plant; the name connotes a discrete, organically connected group of residences. Each toenglu is a spatially contiguous, named area.

In the hills, toenglu are often separated by natural boundaries or by unfarmed areas. There is an average of 7 farmsteads per toenglu in the hills, with a range of 3–20. The residents of a toenglu, at least in the hills where Netting made detailed maps, tend to be partly localized lineages (Netting 1968:146–147).

The toenglu serve only infrequently and informally as a forum for mobilizing labor. The toenglu might be organized for a roof-thatching party, or a small toenglu might join forces with the neighboring toenglu; "occasionally a man may invite a particular neighborhood to work for him, but usually the invitation was an open one, and anyone in the village who is not otherwise occupied may come" (Netting 1968:138).

There are groupings that might be considered toenglu on the plains, but I believe they are too large to be the functional equivalent of hill toenglu, and I discuss them below.

The Village and Supravillage

The village is a fundamental unit of settlement in the Kofyar hills, although the language lacks a term for it. Villages are typically made up of 4–7 toenglu with 25–50 households. The farmsteads are usually adjoining, but some may be scattered, depending on local topography and population. Most hill villages are separated from their neighbors by natural features and are readily discernable on air photos, but some have shared boundaries.

The hill village was a military unit in that the village always fought together, and decisions about whether or not to wage war were made by the village male elders (*daskagam*). It was an economic unit in that labor for mar muos could be drawn from throughout the village. The hill village was a political unit, although villages were not all taxonomically equivalent; some had their own chief, but some recognized another village's chief.

On the plains, villages are contiguous areas of settlement, often lacking physical boundaries, with populations of 31–363. These plains villages are in turn clustered into larger units, a taxonomic level also lacking a name in the local language but which I will call the supravillage, having populations of 265–1503. The supravillages are Kwa, Miket, Lardang, Doka, Doemak, Kwang, and Kwalla.[8] Fogol also might be considered a supravillage. It is now considered a section of Kwa, although it is much more populous than other Kwa villages and probably more populous than the supravillage of Kwang. It now pays its taxes through Miket and formerly paid through Kwang. It also appears to have joined the hill communities in battles (Netting 1974). Temple wrote in 1919 that "though now incorporated the Mikiet and Lardang were formerly independent, and the Mikiet dialect differs from that of the Mirriam" (p. 276).

Supravillages are partly the result of colonial sociosettlement categorizations. Like colonialists elsewhere, the British saw formalized political hierarchies as indispensable for taxing and governing indigenes, and they enhanced or imposed hierarchy within the Kofyar sociosettlement pattern. For instance, the supravillage of Lardang, comprising the hill community of Goetoeng and the plains communities of Kongde and Lumuat did not always function as a decision-making unit before the colonial period. When taxes were imposed by the fledgling colonial government in 1913, the plains portion of Lardang acquiesced, whereas Goetoeng refused (and suffered a vicious punitive patrol that took 13 lives).

The Warfare Alliance

Netting described Kofyar militarism in some detail (Netting 1973, 1974, 1987). He portrayed the Kofyars' historic warfare alliances called *sargwat* ("shield arm"), each consisting of a group of villages that defended each other during hostilities (Netting 1974). The Doka, Plains Merniang, Plains Doemak,[9] and Kwalla sargwat were based on the plains, whereas Jipal and Ganguk were made up of hill communities. The Gankogom sargwat was mainly in the hills but also included the plains subvillages of Lumuat, Kongde, and Kwanoeng (fig. 4.2).[10]

Sargwat are not formal entities. Although some were named, they have no officers, regular meetings, or social functions outside of the armed conflicts that had died out by the 1940s. Their membership was not even completely firm. They were often relative entities, activated by opposition, and not immune to the possibility of internal fission and raiding. As in segmentary lineage coalitions, there was no durable leadership. Thus, even though few young Kofyar adults today could tell you which sargwat their grandfathers fought with, these divisions do reflect social affiliation, or social propinquity, as will be apparent in the analysis of marriage patterns.

The Tribe

Colonial governments in general promote hierarchy in indigenous sociosettlement systems, including both formalization of informal or contested indigenous hierarchies and imposition of hierarchies where none existed before. This hierarchical imperative was especially important in the British colonization of sub-Saharan Africa, with its explicit reliance on indirect rule. Fried (1975) believed "tribes" to be a characteristic product of interaction with colonialists, as is well illustrated by the "mental map of . . . neatly bounded, homogeneous tribes" (Ambler 1988:32) that came to characterize British colonial policy in Africa. Colonial records attest to official satisfaction in cases where "tribal organization had been created" out of "a very disorganized state" (Chanock 1985:112; Berry 1993:28).

It is clear that the "tribe" was the operative unit of social affiliation and classification to colonial British administrators and scholars (Temple 1919), who divided the area shown in figure 4.2 into the tribes of Doemak ("Dimmuk" to the British), Merniang ("Mirriam"), Kwalla ("Kwolla"), Doka ("Jorto"), and Jipal (Ames 1934). The taxonomy was reified by taxation and by the instituting of customary courts, and sociosettlement units considered themselves subordinated to the units that collected their taxes: "In Shendam, where a payment of train ('Zakka') is due from all farmers to their administrative superior, a lessor or claimant to fallow taken up gets half the amount due. (Incidentally this payment sheds an interesting sidelight on the traditional administrative organisation about which so much inconclusive argument goes on. Trace the 'Zakka' and you get the line of subordination.)" (Rowling 1946:29).

Although the colonial tribalization corresponded to indigenous social geography in the case of Jipal, Doka, and Kwalla, it caused problems in the heart of the homeland. In a reflection of the colonial administrators' bias toward the more accessible plains communities, hill villages were classified into tribes named for plains communities and forced to pay taxes through plains chiefs. The chief of Kofyar (who was symbolically the paramount chief) was therefore subordinated to a plains chief, and the Ganguk villages, labeled "hill Dimmuks" villages, were put under a Doemak chief who had held no sway in the hills before (although the administrators were aware of cultural differences between hill and plain and sometimes referred to the hill villagers as a "sub-tribe" [e.g., Josprof n.d.]).

Today, it is by these "tribes" that people from the area often self-identify when they are outside of the area, although the tribes are sociosettlement taxa that correspond poorly to both precolonial political organization and, as shown below, marriage geography. It is somewhat ironic that this level, which is so important by the criterion of self-identification, is in large part a colonial invention.

The sociosettlement unit of "the Kofyar" that I use in this book was defined by Netting on the basis of shared language, agricultural regime, social organization, origin myth, and various other cultural traits, contrasting with those of the neighboring Ron, Mwahavul, and Goemai (Netting 1968:35–43, 1974:146). For lack of a better term, I will call this the "inclusive tribe."[11]

Marriage Geography and Social Propinquity

We have seen that there are multiple levels in the homeland sociosettlement taxonomy and that some aspects of the taxonomy are the same on the plains and the hills while others differ. To discuss the effects of social affiliation on frontier settlement we will clearly need a more distilled breakdown. We can begin the distillation by looking more closely at marriage patterns.

Perhaps the most direct index of social propinquity is marriage patterns, because a marriage not only demonstrates preexisting social propinquity between the spouses' homes, but enhances their subsequent social ties. The bond between *koen*—mother's brother and sister's child—is normally a close one, and koen are expected to feed and look after each other (an oft-described pattern in African societies).

Postmarital residence in the Kofyar homeland is almost exclusively virilocal. None of the sociosettlement units above the farmstead/household is exogamous, and a plurality of most villages' wives comes from within the village (fig. 4.3). The social distance between sociosettlement taxa tends to be quantitative and continuous, rather than abrupt and qualitative: Kofyar are closest to (and most likely to marry) co-villagers, followed by other co-sargwat members, other tribesmen, and other inclusive tribesmen.

The sargwat endogamy in 2,067 pre–1968 marriages is striking, showing a definite tendency for marriages to occur within the same sets of villages that had fought together (table 4.2). Note that the sargwat were not proscriptively endogamous, even in the days when they were active. But the protracted periods of hostility, which sometimes dragged on for decades, amplified the social distance between sargwat, and the postbellum abandonment of lands along sargwat boundaries increased their spatial separation (Netting 1973, 1974).

The relative degrees of propinquity among villages can be depicted by cluster analysis. The symmetrical matrix in figure 4.3 was transformed into a triangular matrix by averaging the row percentages for each pair of villages. This was then used as a proximity matrix for a hierarchical

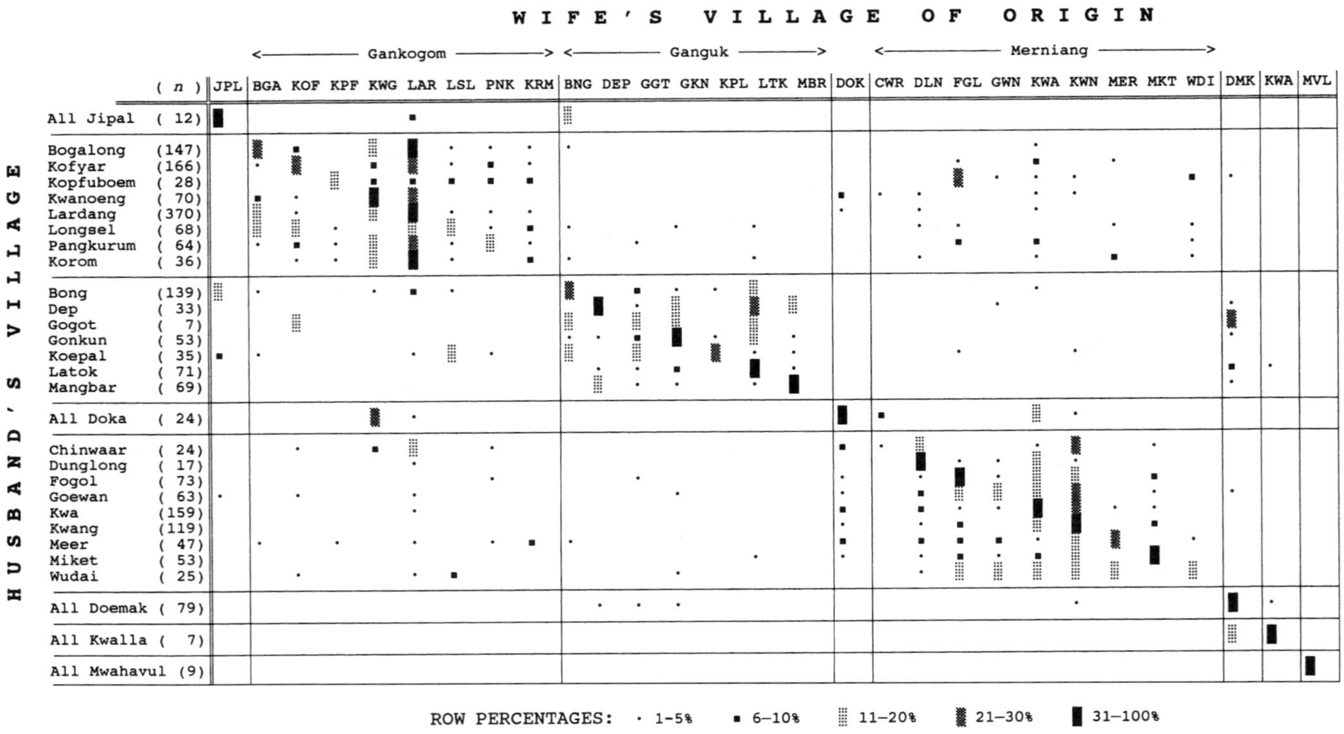

WIFE'S VILLAGE OF ORIGIN

<— Gankogom —> <— Ganguk —> <— Merniang —>

JPL | BGA KOF KPF KWG LAR LSL PNK KRM | BNG DEP GGT GKN KPL LTK MBR | DOK | CWR DLN FGL GWN KWA KWN MER MKT WDI | DMK | KWA | MVL

HUSBAND'S VILLAGE

(n)

All Jipal (12)

Bogalong (147)
Kofyar (166)
Kopfuboem (28)
Kwanoeng (70)
Lardang (370)
Longsel (68)
Pangkurum (64)
Korom (36)

Bong (139)
Dep (33)
Gogot (7)
Gonkun (53)
Koepal (35)
Latok (71)
Mangbar (69)

All Doka (24)

Chinwaar (24)
Dunglong (17)
Fogol (73)
Goewan (63)
Kwa (159)
Kwang (119)
Meer (47)
Miket (53)
Wudai (25)

All Doemak (79)

All Kwalla (7)

All Mwahavul (9)

ROW PERCENTAGES: · 1–5% ▪ 6–10% 11–20% 21–30% 31–100%

Figure 4.3. Marriage patterns between homeland villages. The symbols represent row percentages, i.e., the percent of all wives in the village's row that were born in the column's village. Some categories have been lumped due to small sample size.

Table 4.2 Sargwat Endogamy

Sargwat	Wives in Sample	Percentage from within Sargwat
Mwahavul	9	100%
Doemak	79	90%
Merniang	580	87%
Kwalla	7	86%
Gankogom	949	86%
Ganguk	407	80%
Jipal	12	75%
Doka	24	46%

cluster analysis, using the between-group average linkage method. The resulting dendrogram shows the hierarchy of social propinquity, as reflected in marriage patterns (fig. 4.4). In figure 4.4, the farther the linkage is to the left of the scale, the higher the intermarriage rate, and by inference, the greater the social propinquity. Note that the more common application of hierarchical cluster analysis is the classification of cases on the basis of similarity in their values over a number of variables. Tight clustering in the present analysis indicates not similarity but a high rate in intermarriage, based on the percentages shown in figure 4.3.

The cluster analysis separates villages into four groups. The Mwahavul farmers (the Mwahavul are the Kofyars' neighbors to the north and west on the plateau) in cluster 4 are a clear isolate because there are no intermarriages with Kofyar in the database. There have been a few cases of women marrying into Jipal from Chokfem Mwahavul, but in general an intervening valley acts as a cultural boundary.

Kwalla and Doemak are loosely coupled as another isolate in cluster 2. I have scant data on Kwalla farmers, and I have no doubt that Doemak and Kwalla would be separated from each other if there were more Kwalla censuses. But the separation of the two supravillages from the hill communities accords with the people's characterization of microethnic relationships.

Next largest is cluster 1, containing the villages of the Ganguk sargwat. Latok, Mangbar, Dep, and Gonkun are shown to be a relatively endogamous unit, and in conversation the latter three are often grouped with Latok, a slightly larger village with its own primary school. The close linkage between Bong and the Jipal villages is misleading; the data

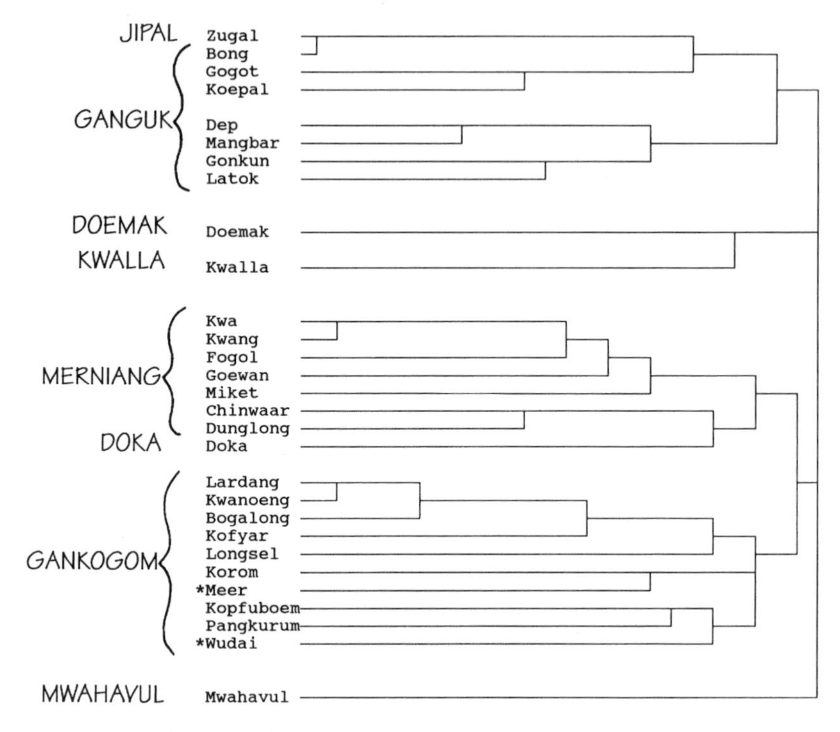

Figure 4.4. Cluster analysis of marriage patterns in the homeland. Asterisks mark the exceptional Merniang village that tends to intermarry with Gankogom villages. Locations of most villages are shown in figure 4.2.

set includes censuses from Bong (into where several Zugal women have married) but none from Jipal (which would have shown considerable sargwat endogamy).

The largest cluster, 3, is the Merniang tribe, clearly split into two subgroups that correspond fairly well to the Gankogom and Plains Merniang sargwat. The sargwat are not totally separated, however, because spatial distance is not completely overruled by the social distance created by sargwat boundaries. For instance, Doka, which constitutes a separate sargwat, is lumped (albeit distantly) with the nearby Kwa infravillages of Dunglong and Chinwaar. This classification also demonstrates how cadastral-cum-political boundaries sometimes contradict the patterns of social affiliation reflected in people's behavior. The plains village of Kwanoeng is as close to Lardang socially as it is physically—the intermarriage rate is high and the villages fought together in the

Gankogom sargwat—even though Kwanoeng has been put under Doka for tax collection.

I have presented several lines of evidence on the geometry of social affiliation among the Kofyar. There is the colonial classification, which was based partly on precolonial Kofyar history and which has influenced postcolonial social affiliation; there are precolonial groupings not reflected in the colonial classification; there is the Kofyars' self-identification; and there are patterns in the geography of the marriages that both reflect and reinforce social propinquity.

The history of mutual defense and the patterns of intermarriage mark the sargwat as a tangible level in the sociosettlement hierarchy, a social taxon with considerable time depth, within which social propinquity is relatively high and between which social propinquity is relatively low. As we turn our attention to the role of social propinquity in settlement locational decisions, this sociosettlement taxon will prove to be important.

5

Frontiering

This chapter describes the physical and cultural aspects of the frontier landscape into which the Kofyar began to migrate at midcentury and examines how and why the movement occurred.

Geology and Ecology of the Frontier

From the base of the Jos Plateau escarpment, the plains slope gradually down to the Benue River. The northern portion of this area, from the escarpment down to a line running from just north of Lafia to just south of Shendam, is the Benue Piedmont. Most of this area is undulating plains formed on Basement Complex granite/gneiss. Inselbergs are common, with large aboveground rock masses predominating in the north and surface rock outcrops toward the south.

South of the piedmont is the Benue Lowlands, geologically situated within a deep trough trending northeast from the Niger River toward the Chad Basin. Marine incursions from the Cretaceous through the Paleocene laid down up to 5500 m of sedimentary strata, predominantly limestones, sandstones, and shales (Bawden and Jones 1972; Offodile 1976). The large erosional plain in the middle section of the trough, sometimes called the Great Plains of Muri (Udo 1970:141), may be divided into the physiographic zones of the Namu Sand Plains and the Jangwa Clay Plains (fig. 5.1). The Kofyar diaspora has occurred across these two zones and the Benue Piedmont; differences in the zones' soils have helped shape the diaspora.

The Benue Piedmont has coarse quartz sand soils formed on weath-

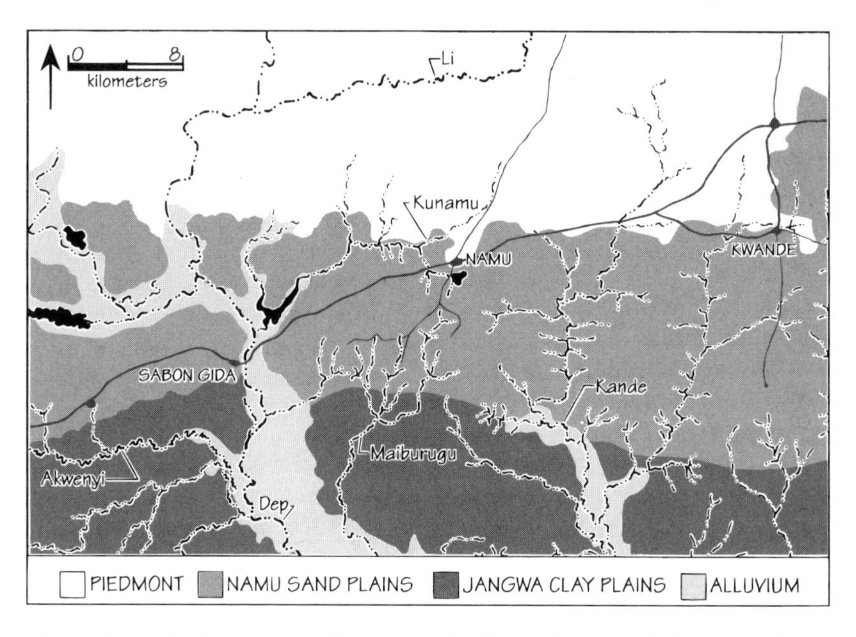

Figure 5.1. Drainage and soil zones in the Benue frontier. Soil information is based largely on Hill 1979.

ered Basement Complex rocks with a admixture of colluvial and alluvial sands from the erosion of the plateau. The piedmont can be further divided into rocky and smooth areas, the former area having more bedrock outcrops and iron pan (laterite) and poorer growing conditions. The old Doemak-Namu road, which runs along the eastern edge of the River Li watershed, separates the zones of rocky and smooth piedmont.

The Namu Sand Plains stretch along the northern edge of the Benue Lowlands, abutting the southern edge of the piedmont. The gently rolling topography is made up mostly of interfluvial crests and upper and middle slopes. The soils are well-drained reddish brown soils formed in Cretaceous sandstones, exhibiting sandy surface horizons.

The Jangwa Clay Plains abut the Namu Plains to the south. Formed mostly in Cretaceous shales, the soils are poorly drained pale clays with heavy texture and strong structure (Hill 1979:16). The land surface is mostly low and flat, with relief between streams and crests rarely exceeding 30 m. There are occasional ridges produced by sandstone bands in the shale, where the soils resemble those of the Namu Sand Plains.

The boundary between the Namu Sand Plains and Jangwa Clay Plains is abrupt in much of the area. It is readily visible on aerial photographs; on the ground, the shift from reddish sands to pale clays is unmistakable except around streams, where it is blurred. The boundary between the Namu Sand Plains and the Benue Piedmont is less distinct but nevertheless definable. Soil characteristics for these zones, described by the Land Resources Development Centre (LRDC) of the British Ministry of Overseas Development (Hill 1979), are shown in table 5.1.[1]

If we define humid tropics as receiving at least 1400 mm of rain with at least seven humid months (Ruthenberg 1980), the Kofyar frontier falls into the wet end of the dry tropics category. The area between Shendam and Lafia receives 1143–1270 mm (44–45 inches) of mean annual rainfall. The rainy season in Namu averages 195 days, and this season increases slightly toward the south. Rains typically last from mid April until late October (Hill 1979:10, maps 4.10a, 4.10b). As one moves from northern to southern Nigeria, annual precipitation increases and also develops a bimodal pattern. The Namu area, close to the geographical center of the country, displays the bimodal pattern with rainfall peaks in June and August, separated by a moderate decrease in July.

The climate in this portion of the middle belt supports cultivation of northern Nigerian crops (millet, sorghum, groundnuts) as well as the southern root crops (yam, cassava, cocoyam) and rice. There is, however, variability in annual rainfall and in the commencement of the rains that may endanger crops. Severe droughts occurred in 1976 and 1983. In 1986 two "false starts" in the rains forced many farmers to plant their millet crops three times. The relationship between precipitation regime and agricultural strategy is detailed in Stone et al. 1990.

This area is within the Southern Guinea Savanna zone. The natural climax vegetation was a woodland dominated by large trees such as *Khaya senegalensis* and *Erthrophleum suaveolens*, with an herbaceous understory of fire-tender species. Primary forest is rare, although Hill (1979:18) reports examples well to the west of the study area; scattered patches visible on aerial photographs of the study area are possible candidates for remnant climax forest as well. The natural climax vegetation was probably reduced by cultivation long before the 20th-century Kofyar arrived on the plains. What Kofyar pioneers encountered was an open woodland savanna with a light canopy, probably dominated by such species as *Daniellia oliveri*, *Prosopis africana*, and *Isoberlinia doka*. Some areas were covered with perennial tussock grasses, especially *Andropogon* and *Hyparrhenia*, often more than 3 m in height.

History of the Namu Plains

Little information is available on the history of Namu District and this area of the Benue Lowlands in general. Colonial period documents now housed in the National Archives at Kaduna (NAK) contain some information on economy and settlement in the area prior to the Kofyar entrance, such as Rowling's 1946 report on land tenure. Other writings concerning Namu have been collections of oral legends (Dagum n.d.) and conjectures (e.g., Agi 1982; Adamu 1978).

The Muri Plains have contained vast expanses of uninhabited territory throughout this century, largely due to the dangers of raiding. The threat came first from the Jukun empire to the south and later from Muslim emirates to the east and west. The Jukun empire dominated the area in the 16th and 17th centuries (Fremantle 1922:32–33, 43; Ames 1934:27–28). They extracted tribute from the Goemai (Ankwe), who in turn raided the Kofyar for slaves. Oral histories agree that Jukun influence north of the Benue River declined in the late 18th century, but the Fulani jihad of the early 19th century left the plateau tribes ringed by dynasties, "all militant with the enthusiasm of their youth" (Ames 1934:30) and wont to prey on plateau groups in accessible areas.

An emirate established at Wase (northeast of Shendam; see fig. 4.1) as an outpost of the Bauchi dynasty began to threaten the plains in the early 19th century (Fremantle 1922:6). By the late 19th century, the Goemai area was said to be the best source of slaves for the Lafia emirate (Hogben and Kirk-Greene 1966:548). Yet a small population remained on the Muri Plains despite the hazards. Southeast of the Kofyar were the Goemai towns of Shendam, Kurgwi, and Kwande, all believed to be quite old. Farther west were the smaller settlements of Njak, Shindai, and Namu, which bore cultural characteristics of both the Kofyar and the Goemai.[2] Kurgwi, Kwande, and Namu, which had defensive earthen walls, were sustained by shifting cultivation on lands beyond their walls.[3] They were still vulnerable to outside attacks, as in 1820 when Yakubu of Bauchi reportedly "over-ran the country of the Ankwe and reduced all their walled and stockaded towns" (Fremantle 1922:44). Denizens of the Njak area sometimes fled to the hills for refuge from slave raiding (JOHLT 1981:641), and Namu was harassed by raiders coming east from Lafia (Findlay 1945:139). Pang Matlong, the inselberg on the northern edge of Namu, may have provided shelter as well. Kofyar legend relates that the culture hero Tuupyil, brother of Paya, created the inselberg as a refuge against slavers.

Table 5.1 Characteristics of Land Systems in the Study Area

	Piedmont (Rocky)	Piedmont (Smooth)	Namu Sand Plains	Jangwa Clay Plains
LRDC Land System Number	587	588	567–582	572
Geology	basement	basement	sandstone	shale
Dominant Slope	5.2%	5.2%	3.5%–4.4%	1.7%
Drainage Texture	medium–very fine	fine–very fine	medium-coarse–medium-fine	medium–coarse
Soils				
Drainage	well	well	well	imperfect–well
Surface Texture	coarse	coarse	coarse	fine
Exchangeable Potassium	0.1	0.1–0.14	0.1	0.16–0.2
pH	5.2–5.7	5.5–5.8	4.6–6.3	4.7–5.6
Soil Limitations[a]				
Yam	t	t	t	Ct
Millet	c	c	—	DCt
Sorghum	tc	tc	—	dct
Rice	*D*T	*D*T	*D*TC	*DC*
Groundnut	c	c	—	ct

Source: Hill 1979.

	Texture	Drainage	Coarse Material
[a]Severe	*T*	*D*	*C*
Moderate	T	D	C
Minor	t	d	c

Threats also came from local groups. An early British patrol in Ankwe District was chased off by "marauding Montols" (Fremantle 1922:10), who lived in the hills east of the Kofyar and probably also on the adjacent plains. The Montols even claimed to have repulsed Fulani attacks in the preceding decade (Fremantle 1922:46) and were among those raiding trade routes such as the Ibi-Wase road during the early 1900s.[4]

But having quelled these dangers, the British were puzzled and frustrated by the Kofyars' reluctance to move into this area, "where they could finally find enough food to make prosperous tribes and where they could be easily controlled by the government" (Netting 1968:193). Findlay, who had obviously spent time in the picturesque hills, speculated whether the Kofyar were "moved by anything in the nature of a sub-conscious aesthetic appreciation" (1945:139); a 1955 report opined that few Kofyar had moved to the frontier, "and it is to be hoped that they will not find it necessary to do so: who would live on a plain if he could live on a hill?" (quoted in Netting 1968:195).

Causes of Population Movement

The Kofyar migration is one of a series of "downhill movements" of populations out of hilly areas of Nigeria in the mid 20th century (Conant 1962; Gleave 1965, 1966; Udo 1966; cf. Gleave 1963). Many of these population movements, including the Kofyar case, were facilitated by the Pax Britannica, which reduced or eliminated the threat of raiding and slaving on lowland areas. The Pax was a necessary but not sufficient condition for hill abandonment, and Gleave (1966) cites a series of "push" and "pull" factors promoting movement:

PUSH	PULL
pressure on cultivable land	availability of abundant land
lack of building land	need for cash
disease and malnutrition	outside contact
government action	government action
inaccessibility	

Gleave cites tradition, tribal custom, and disease as militating against movement.

Land pressure was without doubt a fundamental impetus for the Kofyar to move to the frontier. Intensive agriculture, especially in the dense plains communities, offered low marginal returns to labor. The plains around Namu were practically empty; the population density for all of Shendam District was estimated at 13/km^2, or 34/mi^2, in 1952 (Grove 1952:52). My analysis of 1963 aerial photographs yielded a population density of 10/km^2 for the Kofyar settlement area itself, and this was surrounded by unpopulated savanna.

At times, the homeland food base was simply insufficient. There was sporadic intervillage food raiding in the earlier part of the century, triggered by agricultural shortfalls (Netting 1974). The usual pattern was attacks by crowded plains villages on the better-stocked larders of hill farmers, but hill raiders came down the escarpment on occasion as well.[5]

Whereas these altercations probably resulted from acute, severe food shortages, there were milder but chronic food shortages in the areas with the highest population density. Population density correlates strongly with migration rates, as measured by Netting in the 1960s (Stone et al. 1984). The migrants Netting interviewed in 1966 almost unanimously cited the need for more food as a factor in their movement to the plains. The frontier offered not only relief of the crowding in parts of the homeland, but dramatic boosts in marginal returns to labor, because the pioneers promptly abandoned their painstaking intensive techniques for swidden tactics.

After the population density differential, the most important catalyst for frontiering was to earn a profit on the agricultural market, although the "need for cash" that Gleave cites is misleading. The Kofyar economy had been very slightly monetarized at least since the early 20th century. A market had operated in Kwalla since before the arrival of the British; Kwalla farmers were reported in 1912 (three years after their first contact with whites) to have "taken readily to trade," and a 1927 report (Glassan 1927:4) mentions Kofyar marketing palm oil, woven mats, and agricultural implements. A market established in Doemak village in 1937 eclipsed the older Kwalla market (Findlay 1945:140).

The Plains Merniang area was poorer than that of the Plains Doemak in the palms that produced the highly marketable oil, and Merniang farmers earned cash by headloading yams, salt, and palm oil up to the plateau mining areas.[6] Merniang participation in overland trade capitalized on the absence of a direct road from the Shendam area up to the mines; what was to the Kofyar a 30 to 40-mile trek on foot was a 130-mile drive. An all-season road finally connected Panyam and the high plateau with Shendam in the late 1950s, reducing the incentive to transport local goods at the same time that it bettered the agricultural market on the frontier.

Yet it is easy to overestimate the role of overland trading in the Kofyar economy. Virtually all Kofyar were full-time subsistence farmers during the agricultural season, and the percentage of men who made yearly trips to the mines was fairly small. There are stories of men who made a single trip in their lives, using the proceeds to purchase a loincloth and a hoe.

Before the colonization of the Namu area, wealth took the form of such commodities as domestic animals, salt, and iron hoes (Glassan 1927:4; Netting 1968). Hill communities in particular, where the economy was even less monetarized than on the plains and where per capita farm production was higher, converted surpluses into expensive stock such as horses, dwarf cattle, and, by the 1960s, pigs.

Taxation by the colonial government was an inducement to obtain cash, although taxes were initially payable in kind (JOHLT 1981:642). Collection of taxes through the native authority was as much an instrument of subordination as of raising funds, and delinquent villages could have livestock seized for taxes.[7]

The head tax in Doemak and Merniang was 3 shillings per adult male as of 1927 (Glassan 1927:5), rising to 2 Nigerian pounds in 1961 (Netting 1968:165). Nevertheless it is doubtful that taxation was an important "pull" factor in the Kofyar migration. The Shindai area, which was taxed at the higher rate of 5 shillings (JOHLT 1981:630), contributed little to the migratory stream. But tax demands did affect where the migrants ended up. One reason for shifting from lands east of Kwa to south of Namu was taxation. Merniang farmers were cropping an expanding area around Zomo, and the chief of Shendam was demanding tax from them.

Overall it was less the need for tax money than the desire to voluntarily enter the market economy that spurred the Kofyar migration; the distinction may not be essential for understanding settlement, but it is important for understanding indigenous agricultural development (Netting et al. 1989, 1993). Once on the frontier, the Kofyar entrance into the market was swift, with households that controlled only minuscule amounts of cash in the homeland bringing in a median of well over 50 Nigerian pounds after three years on the frontier (Stone et al. 1984). Two marked effects of colonial rule had been the improvement of the transport infrastructure and stimulation of urban growth (Hart

1982:44–45), both of which enhanced the marketability of Kofyar agricultural surpluses.

Several of Gleave's other push and pull factors were absent in the Kofyar case. The Kofyar case is valuable for modeling settlement processes independently of government policies because, unlike so many other African cases (Udo 1966; Gleave 1965; see Morgan 1983:109–111), with the Kofyar the colonial government had only a minor impact on the movement and virtually none on the specifics of settlement pattern. The government was obviously responsible for the suppression of military threat and was in general encouraging the movement of farmers to the empty plains, where they could better produce a surplus and have it taxed. But in the final analysis, population flowed onto the frontier at a pace selected by the Kofyar, and Kofyar decisions on where, when, and how to settle were independent of government intervention.

The government did forcibly remove nine hill villages to the plains in 1930 after an assistant district officer was killed in a dispute in Latok (Netting 1987). Villagers from the hills above Doemak were resettled near the river east of Kwalla and in the piedmont between Doemak and Njak. The refugees repeatedly attempted to return to their homes until they were permitted to do so in 1939.

If the forced removal to the plains had any effect on the later migration it was probably to hinder it. The Latok area was the last to join in the later movement south, largely because of the relatively low land pressure (Stone et al. 1984) but perhaps also in part because of the grim experience of the 1930s. The colonial officers saw the "Hill Dimmuks" as motivated by "the spiritual value of their former homes, in which the graves of their ancestors and other objects of veneration are situated" (PRO 1939), but there must also have been a compelling interest in protecting the valuable palm trees. When the voluntary migrations began soon after the Latok incident, these homeland investments were always maintained by a resident portion of each household.

There is no evidence that disease played a role in the frontier migration. Although trypanosomiasis was present in the hills in the decades preceding the movement (Glassan 1927) and may have been exacerbated by the development of uninhabited buffer zones separating warfare alliances in the hills (Netting 1973), the problem had abated by the time the migration began. Moreover, large tracts of the frontier were also infested with tsetse until bush clearing gradually reclaimed the land from the disease vector.

History of Migration

In the homeland, farmers from both hill and plains villages had been cultivating bush plots (*wang*) since well before the diaspora began. In fact, one of the earliest written accounts of the Kofyar notes, "Of late years, the Kwolla, Dimmuk, and Merniang peoples have been paying small amounts, in grain, to the Ankwe people in respect of farms made by tribesmen on land which the Ankwe claimed as belonging to them" (Fitzpatrick 1910:51). Some plots were far enough from home for the farmer to build a field house to stay in for part of the growing season. Some field houses turned into residential compounds when land pressure was especially high, and some were abandoned when population ebbed.

The spread of permanent settlement was checked by local soils. The Plains Merniang, bounded to the west by the plateau escarpment, looked to the east for expansion. But land to the east was patchy; flows of vesicular basalt were widespread, and the basaltic areas were labor-expensive to cultivate. The sandy soils within the basalt flows were already occupied by the sprawling settlement of Kwang.

The Plains Doemak, with their backs to the escarpment, faced more than 25 km of rocky piedmont to the south before the beginning of the Namu Sand Plains. Large tracts of the piedmont here were uncultivable because of outcrops of basement rocks or heavy laterization. Other areas scattered through the piedmont were more favorable. I have only circumstantial evidence on the priority of settlement location criteria for these early pioneers, but 1963 aerial photographs show that agricultural plots favor areas of low relief near streams or rivers, which tended to be free of outcrops and laterization.

The Kwalla, by contrast, could spread in several directions, especially to the east and south, where there were broad areas of smooth piedmont and alluvial soils. Kwalla farmers led the agricultural colonization to the south.

Although Findlay was only vaguely aware of these variations in agricultural potential, he described the first rumblings of the agricultural diaspora:

> Unfortunately, now that the pax Britannica allows free movement, there is an increasing tendency to neglect these home farms and to adopt shifting cultivation in the bush to the south. The movement has been adopted in varying degrees by all the tribes in this thickly populated region. Dimmuk's [Doemak's] eastern neighbors, the Kwolla [Kwalla], were the first to start,

> and, being on the perimeter of the thickly populated area, they had a free field before them. In their scramble for new and easily cultivated land they too, are gradually abandoning the intensive cultivation of home farms. . . . Already farms spread to distances of twelve miles or more, and some 100 square miles of bush have been affected. The people have established farm settlements which were at first occupied in the farming season only, but in which some members of the family now live throughout the year. . . . The process has not been quite so easy for the Dimmuk. To north, east and west expansion is barred by populated country or rugged hills, and there is a good deal of rocky country immediately below them; consequently, they have farther to go in their search for new land. (1945:139–40)

The following year, Rowling reported (1946:29) that the Kwalla and Doemak had farms as far east as Shendam in land that was virgin ten years before; furthermore, "Dimmuk are approaching Kwande and are back round Dungban: and there are several recent Tiv settlements between Lamkaku [Langkaku] and Kwande." Merniang farmers had apparently skipped over the basalt flows and were farming east of the Shemankar River (Rowling 1946:14).

Rowling compared the Kofyar migrant farms to those of groups originating from more distant areas on the plateau, whose longer migrations had led them to establish more permanent settlements, and to the Tiv, who had long been in the practice of shifting settlement increasingly north of their homeland to the south:

> Mirriam Kwolla and Dimmuk have not reached the Eggon Montol Yergam Tiv Jarawa or Bi Rom [Birom] stage of permanent settlement with, in at least some of these cases, burial of their dead at their new homes. The former groups establish a second compound on their new farms, but their roots are still in the parent village where they spend most of the dry season and maintain their farms during part of the rains. The bulk of their crop from their new farms is carried home, though the Dimmuk at least, with unfortunate results to their old field, are tending to leave most of their livestock in charge of a wife [sic] or wives at their new establishments. . . . Sons however often clear each his own land in the Njak, Dungban, or Kurgwi/Kwande bush, or take up a piece of Kwolla fallow which the Kwolla farmer has left in his push southeast. Only a few Dimmuk have asked to be registered in the Namu tax lists, with the implication that they have settled permanently; but the time will soon come when they will get tired of carrying their crops 12 to 15 miles. (1946:14; punctuation as in original)

The majority of farmsteads erected in the early stages of the diaspora were field houses, or satellites to home farms. In most areas the Kofyar were practicing shifting settlement with fairly short occupation spans. The settlement histories record 62 farms established outside of the homeland before the early 1950s, 49 of which were in the Njak-Shendam-Kwalla bush area (fig. 4.1). More than half of these settlements were abandoned in less than 10 years, and more than 90% were left within 20 years. These residence times are much shorter than on the frontier south of Namu, as I describe in chapter 11.

Grove (1952:55) described the shifting agriculture south of the plateau at this time. Groundnuts were being planted in the first year, followed by sorghum or risga. Within 5–6 years, the humic topsoil was said to be gone, yields dropped, and fields were invaded by the purple-flowered parasitic weed *batawata* (*Striga hermonthica)*; then the farmers moved on.

By the mid 1940s, Plains Doemak were farming the bush south of Shindai and probably by the late 1940s had moved into the area northeast of Pandam where the rocky piedmont gives way to the Namu Sand Plains. Some Merniang farmers were there as well, probably including Doedel, the Kwa chief who later pioneered on the plains south of Namu. By the late 1940s, the search for bush fields had reached as far south as Namu and Kwande. If colonization was to continue, it would have to leapfrog to the south over the Namu and Kwande bush fields.

Early Frontiering on the Plains of Muri

In its early stages, the Kofyar diaspora had been a gradual affair, with farms being periodically abandoned for spots farther out on the piedmont and with some hopscotching from area to area. But the movement onto the Namu Sand Plains was a major step. There was first the matter of distance. Most of the bush plots the Kofyar had worked in the 1940s had been within 20 km; leapfrogging over Namu's cultivated perimeter left some of them 60 km from home.

But the land beyond Namu was almost empty, with only a light scattering of Tiv practicing a mixed farming/hunting/fishing strategy. Throughout the colonial period, the Tiv had followed a pattern of migratory expansion that led them out of their homeland near the Benue River. By the early 1940s some Tiv were farming along the trail coming east from Lafia. In 1946 Rowling mentioned (1946:28) recent Tiv settlements between Langkaku and Kwande, and the present Kofyar communities of Koprume, Kopdogo, and Duwe are said to be named for earlier Tiv inhabitants.

The south country was also untamed, with a denser tree canopy and an abundance of wildlife. Elephants endangered people and trampled crops. The Muri Plains south of Shendam had been described a few years before: "[T]he country slopes very gently down towards the River Benue, with an increasing quantity of trees and bush until it becomes a vast forest, thick and sunless, except for the open space of its many swamps, uninhabited by man but the home of the bush cow, antelope of many kinds, an occasional lion, and the tsetse fly, and visited annually by elephant" (Ames 1934:11). But the frontier beyond Namu offered some enticing dividends to pioneers. Unlike the piedmont soils in Basement Complex granites, the soils beyond Namu were formed in sedimentary parent material. Their initial productivity was high, but more important was an intensification slope that in most areas lacked the sharp drop-off of the piedmont soils. High yields could be sustained for long periods, and even then, the soil responded to efforts to boost production concentration (examined further in chapter 11).

The turning point in the march into the Benue savanna was approximately 1951, when significant numbers of pioneers began farming near Namu and across the Namu-Kwande road on the Namu Sand Plains to the south. The movement occurred on several fronts. The Kofyar community that began to form just to the northwest of Namu comprised mostly Muslims. It was led by Mallam Musa Gokla (originally from Fogol), an early Muslim convert. At approximately the same time, Plains Doemak farmers crossed the road west of Namu into Koprume; they also settled south of Kwande in an area they called Mangkogom. The diaspora was encouraged by Felix Dakyap, the Doemak chief, who by 1959 had succeeded in getting the Namu-Doemak trail completed.

Meanwhile, Wubang of Fogol, a leader of the Zomo community that was being dunned by the Shendam chief, approached Ali, the chief of Namu, and was directed to the empty savanna south of Namu's fields. Doedel, the chief of Kwa, was a driving force behind movement into this area. Ungwa Long ("ungwa of the chief") was named for him, and the Kofyar still recall his campaign to convince others to join the pioneer community. He argued that the move could relieve the pressure on homeland resources and also allow the Kofyar to enter the market. Joining Doedel and Wubang were Damulak (the chief of Lardang) and several other farmers coming mostly from the Plains Merniang. This was the real origin point for the settlement system I came to study.

The early pioneers tended to come from the largest households

(Stone et al. 1984). Because no Kofyar were ready to abandon their home farms, bush farming favored households large enough to staff two farms. Many households started on the frontier by sending down a fragment of the household, often the head, a wife, and a son working for the growing season. Doedel, for example, stationed several brothers and wives on the bush farm, and he traveled between home and the frontier on a horse.

6

Pioneer Agrarian Settlement

Although Kofyar bush farm settlements had been creeping outward from the homeland throughout the 1940s, the movement to the Namu Plains that began about 1951 was in many ways a fresh start. The Namu Plains settlers had jumped over the piedmont north of Namu, with its thin soils dotted with bush farms. The savanna where they landed was different in several key respects. It was almost empty, with minimal constraints on settlement decisions. Its sedimentary soils offered a dramatic improvement over the productive capacity on the piedmont and began to act as a real stimulus to the growth of the new settlement. It was significantly farther from the homeland than previous Merniang bush farms at Zomo and Doemak farms around Kopdayim—enough to make a real difference in commuting. Pioneers on the Namu Plains were doing more than planting outfield plots; they were migrating.

I have noted "how fascinating it would be to monitor the evolution of a real agrarian settlement system, beginning at a 'zero point' with a small initial pioneering population" (Stone 1991b); this was, in many ways, just that. In this chapter I look more closely at the development of the settlement pattern, drawing on interviews, censuses, air photos, and ground survey (see the appendix for further description of the research instruments used in this study). After sketching the early agricultural strategies, I first focus on the nature of the frontier "settlement" (referring to the farm compound) and factors resulting in dispersal or agglomeration at the level of the individual settlement. I then describe the sociosettlement construct called the *ungwa*. This chapter will not be the final word on individual settlements or ungwa, but it is necessary to introduce these levels of settlement and some of their history before looking more analytically at them.

Early Agriculture

Kofyar agriculture on the frontier differed in several respects from homeland farming. The homeland staples of millet, sorghum, and cowpeas were mainstays on the frontier too, and secondary crops such as peanuts and maize were grown as well. But new crops also came to be very important: yams and rice.

The Kofyar began to cultivate frontier plots by pulling the tall grasses that covered much of the savanna. After this, they killed trees individually with small fires around the trunk. Most of the functions of the general burn were unnecessary here: the soil was highly fertile without tree ash, and the insect predator load was light. Left standing, the dead trees served as stakes for yam vines to climb and without leaves did not block sunlight.

It was common to begin the frontier farm with a crop of sesame. This was a labor-cheap crop that could be started during the October pause in the homeland schedule, when some farmers moved between homeland and bush farms. Sesame was followed by either intercropped millet-sorghum or yams. Rice was grown in small quantities in streambeds.

I have no measures of the concentration of inputs, but it appears that the agricultural regime was intermediate within the range of extensive agriculture described in chapter 3. The general character of farmwork was industrious and highly energized, as reflected in Netting's first description of a frontier work party: "The work was accompanied by two drummers. Each person did about 8 feet on three parallel rows, working at top speed, and then ran to the next place shouting" (Netting unpublished field notes 1961). The homeland's requirement of incessant farmwork had long since been instilled into the Kofyar ethic, and it was not dislodged by this move to a frontier.

Unlike highly extensive farming, the crops were planted in a thoroughly worked field, the grains going into parallel ridges and the yams into knee-high heaps. The crops were weeded, the grains receiving two weedings and the yam heaps being scraped several times during the season.

Still, the time invested in weeding was low by intensive standards, and in other ways farming was markedly more extensive than in the homeland. The close waffle ridges in which seeds were planted in the homeland were replaced by widely spaced parallel ridges. The homeland practice of annually fertilizing fields with goat-dung compost was unnecessary, which saved not just the work of spreading compost but the work of tending tethered goats throughout the farming season. Often 10 ha in size, frontier farms were big enough to allow the farmer to abandon one plot when yields dropped, switching cultivation to another part of the farm. Gone too were the time-consuming tasks like making nursery beds, transplanting seedlings, constructing intricate waffle ridges, and farming outlying terraces.

Rather than the small-scale, continuous (Netting 1965) labor demands of the homeland intensive system, agricultural work on the frontier had a few bottlenecks separated by periods of lighter demands. This sort of labor profile is characteristic of swidden extensive farming, especially in savanna environments (Burnham 1980; Richards 1983), even if the actual labor demands differed from those in classic slash-and-burn systems. For instance, the early-season bottleneck was caused not by felling, drying, and burning trees en masse, but by hoeing ridges for grain and heaps for yams.

Early Settlements

The Namu Plains pioneers had few constraints on settlement form. Land was cheap and abundant, and external meddling in the settlement process was negligible. Kofyar pioneers were at liberty to replicate the dispersed settlement pattern of their homeland, form concentrated settlements like other plains groups, or develop alternative settlement forms.

The first Kofyar occupation south of Namu, with a small seasonal population burning trees in an expanse of savanna populated mainly by baboons and elephants, had a pioneer atmosphere that is reflected in Netting's description: "After taking a fork in the road, the road became little more than a bicycle track. . . . [O]n three miles of trail we passed 41 homesteads. Those at the end of the line had been there three years and were still engaged in burning trees. Their homesteads had a pioneer look, with a scattering of rather rude buildings, a drying rack or two, and the circular mar [millet] drying stand with mud between the uprights to make a cooking shelter" (unpublished field notes 1961).

Figure 6.1 shows 1963 settlement and land use in what I am calling

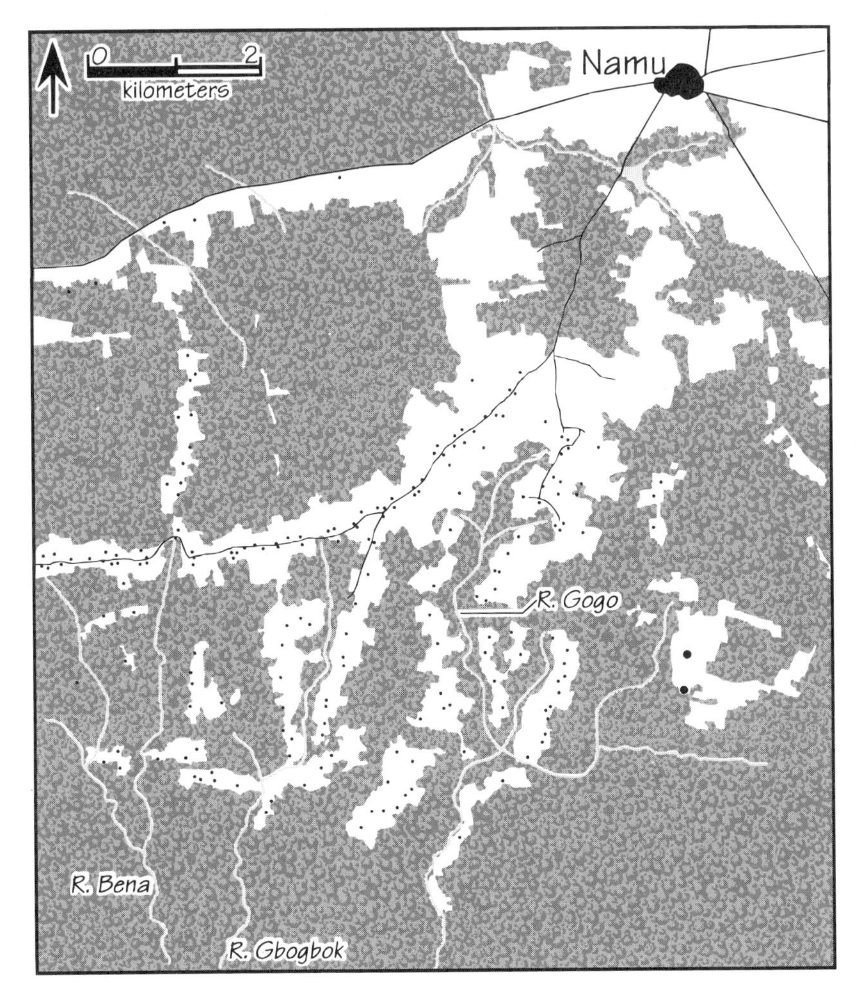

Figure 6.1. Settlement and land use on the early frontier, 1963. Based on analysis of aerial photographs, augmented by surveys and interviews in 1984–85. Small dots are Kofyar compounds; large dots are Tiv compounds.

the core area.[1] Kofyar settlers made a point of occupying a generally contiguous area, in part because of animal predations. Padoch (1986:283) describes precisely the same phenomenon for the Lun Dayek in Sarawak, who "like many other cultivators also attempt to locate new farms close to other farms being worked in the same season; a wide scattering of fields increases the labor required in protecting crops from the depredations of sparrows, monkeys, and other animals."[2]

Contiguous settlement also serves to maintain a perimeter against other cultural groups. Although settlement in its early stages was not in danger of incursion by other groups, there were Tiv in the area, and because land was being colonized rapidly, the future advantages of controlling contiguous areas was obvious. In recent years attempts to maintain perimeters have led to conflicts between Kofyar and Tiv (Stone 1994a).

What does not show up readily in figure 6.1 is that the ubiquitous homeland pattern of spatially discrete compounds housing independent households gave way to a more varied pattern that included different types of minor aggregation. These aggregations were created by both accretion and design. The first settlement in the core area was definitely more agglomerated than homeland settlements had been. In 1952 the group—including the chiefs of Kwa, Fogol, and Lardang—began building a tight cluster of 4 compounds on a spot approximately 4 km down the trail from Namu and 700 m west of Rafin Gogo (see fig. 10.1). These pioneering households were large by Kofyar standards; one man reported rotating nine wives and two adult sons between home and the bush farm. The residential grouping was further expanded by pioneers who stayed there for a few months or years before dispersing to reside on their own plots. Of the 99 pioneer households censused in 1961, 17 first resided in this cluster. By the mid 1950s, there were an estimated 30–50 people occupying this settlement, which I call a macrocompound.[3]

The macrocompound fissioned at about 1955, producing a planned macrocompound. A group of households, joined by several others just coming to the frontier, broke away to form a macrocompound of their own, comprising approximately a dozen households. They moved across to the unoccupied eastern side of Rafin Gogo. Whereas the chiefs' macrocompound had grown by accretion, the new compound was a planned multihousehold settlement from the beginning, probably with a population in the 30s. Residents called the settlement a "company," or "Goewan *gari*," using the Hausa word for a town.

Residents of the company point out that the east bank of Rafin Gogo was obviously going to be a prime area for agricultural settlement. An old frontiersman who had helped found Goewan gari told me that they wanted to dominate the area by establishing the first settlement of size and duration there—echoing the geographers' concept of the first viable settlement (Jordan and Kaups 1989). They were generally successful in this, as shown in the discussion of ungwa below.

Another macrocompound had appeared some 2.5 km south of the original settlement by 1963, comprising 5 households from the hill village of Pangkurum, who migrated as a group. The 1963 aerials show the macrocompound and dispersed fields pattern in what is now the Pangkurum section of Wunze (described in chapter 8).

There were more modest forms of settlement concentration as well. Of 99 bush farmers censused by Netting in 1962, 49 said they had first lived in other frontier compounds, for periods ranging up to several years. In many cases this was simply a matter of several households residing in a compound together when the farms were first starting. There were also compound pairs, usually with the compounds on either side of the path and their farms reaching out behind them. Several of these pairs are visible on 1963 air photos, and the farmers in these settlement dyads often worked collaboratively. Settlement pairs can be seen today in newly settled areas. I visited one south of Sabon Gida; the two young families had recently erected compounds about 30 m from each other, and the families worked together in what they called a joint effort.

Case Study of a Pioneer in Ungwa Goewan

Boyi Pankok is a small man, with white hair and a pair of glasses that gives him a scholarly look. Now in his 70s, he spends much of his time resting, although he is still a *wupinlu* (household head) and still operates a farm—in fact, he grew 5000 yams the year we were there. Boyi was one of the first to take up farming on the frontier, coming from his home in Goewan Kwa to farm in what is now called Ungwa Goewan.

With several friends from Goewan, Boyi had come to begin working a *wang* (bush farm) at about 1954, two years after the group headed by the chiefs had arrived. The new arrivals moved into huts in the chiefs' compounds, which were quickly accreting into a macrocompound. Because the area around the macrocompound was being rapidly claimed, the Goewan farmers claimed land on the east side of Rafin Gogo. Each erected a small field house on his property to provide shelter from the rain, while still residing in the chiefs' macrocompound.

By the second year, there was a sizeable contingent of farmers ready to put up permanent residences east of the stream, including those that had already planted there plus several others from the homeland. All were from Goewan, and they decided as a group to erect a composite residence, or macrocompound, near the northern end of what is now Ungwa Goewan. The settlement had 15–20 huts, housing approximately a dozen farmers and some of their wives. They called it Goewan gari, in a whimsical use of the Hausa term for a town or city.

However, Goewan gari began to disintegrate within four or five years as farmers were pulled to their plots. Boyi moved into a small compound on his land after four years of commuting from the gari. He planted the land immediately surrounding his new compound and farmed it for approximately eight years—long enough for the fields to be encroached by *Imperata* grass—and then razed the huts and shifted his residence to a new part of the farm.[4] This spot too was farmed for approximately eight years, until he built a third compound on his present location. Claiming to be too "tired" to move again, Boyi intends to remain in his present compound until he dies or "retires" to his old home in Goewan, Kwa.

Boyi's residential shifts within his farm averaged eight years. Although I have no way of knowing the size of Boyi's cultivated area, Netting's measurements of 10 frontier farms in 1961 yielded an average cultivated area of 10.7 acres or 4.3 ha (Netting 1968:200). Original, unfragmented farms in this area were typically more than 6 ha in size and sometimes more than 10 ha, so it is a good guess that he was working around a third of his land during each eight-year period before moving on to virgin or fallowed land. This is clearly the classic combination of shifting settlement with relatively extensive agriculture, but it was taking place within a bounded area over which the farmer had ownership rights. Like most of his neighbors, Boyi made agrarian settlement decisions in anticipation of the day when he would be farming his land, or a fragment of it, on a much more intensive basis.

Factors Shaping Early Dispersion and Agglomeration

There were several related reasons for the different types of small aggregation. One factor was that the pioneering group included many fragmentary households, often a household head with one wife and perhaps an adult son. Hard-pressed to erect an independent compound and keep house while they were trying to get a farm started, these units often joined into composite households. The massing of multiple households to keep domestic work from eating into field work was precisely what was observed in new settlements among the Kekchi Maya, whose pioneer settlements were in the same size range as Goewan gari (Wilk 1988:138, 1991; see also Keegan and Machlachlan 1989).

Another reason for settlement concentration was defense. This is a common stimulus for settlement concentration, although in this case the threat came from wild animals rather than people. Larger domestic groups meant more protection against baboons, who were wont to make off with chickens; elephants posed a threat as well, and at least one farmer was killed by a buffalo.[5] However, as the savanna was gradually replaced by fields filled with crunchy seed heads of millet and sorghum, wildlife posed more of a threat to crops than to the domicile, promoting dispersal.

Impetus for early settlement concentration also came from agricultural labor, and the relationship between Kofyar agricultural labor and settlement accords with Trigger's (1968:62) observation that relations of production are often given physical representation in the spatial disposition. If large household labor pools were valuable in meeting the demands of frontier farming, they were almost essential to the simultaneous running of homeland and frontier farms. The effects of the change in labor demands were unmistakable: frontier farming households had 26% more adult workers than traditional households; the population of early colonists was biased toward households that were large even before the migration (Stone et al. 1984); and households steadily increased in size on the frontier, from a premigration average of just over four to an average of more than nine after a decade on the frontier (G. D. Stone 1988:162; Netting et al. 1993:221).

But the larger co-resident groups were also augmented by suprahousehold labor, as there was increased reliance on the cooperative labor groups of *mar muos* and *wuk*. Whereas mar muos had mainly worked outlying plots in the homeland, they were convened on households' main fields on the frontier.

If large labor pools were adapted to frontier agriculture in general, they were especially advantageous in the pioneering phase, because of the labor demands of field clearing and hut building and the fact that the farmers were only on the frontier temporarily, returning to their home farms in the dry season (and sometimes even during the growing season). At the same time, the low level of agricultural intensity placed little premium on minimizing the residence-to-plot distance. The suprahousehold compound was therefore well adapted to early frontier farming.

It is no accident that the Kofyar macrocompounds resembled the settlement form of the Tiv, who also practiced shifting cultivation on a generally communal basis and who would have faced similar agricultural labor demands (Bohannan 1954a). The Tiv reside in multiple-family households that may have populations well over 100 (Stone 1993a).

The changes from the homeland settlement pattern on the frontier are intriguing. It is undoubtedly common for frontier migrants to prefer replication of their homeland settlement, whether the homeland pattern was agglomerated or dispersed. Udo (1966:135–136) describes Yoruba settlement as maintaining its nucleated pattern during moves from hills to plains; Conant (1962) gives an example of dispersion being maintained in a similar move. Yet immigrants also show a consistent willingness to replace their preferred settlement patterns to adjust to changes in ecology. In Nigeria, many cases of downhill movement brought a change from relative agglomeration to dispersion, including the Ibo (Udo 1966:136) and numerous Jos Plateau groups (Gleave 1965, 1966). By the same token, Russian immigrants in Kansas and Ukrainian immigrants in Saskatchewan, who began by recreating their old agricultural villages, soon adopted individual holdings (McQuillan 1978; Friesen 1977). In the colonial United States, there was a pattern of failure of the clustered rural settlements that replicated Old World forms, the result of both ecological and social causes (Jordan 1976). What was less clear is the likelihood of immigrants from a dispersed settlement pattern adopting agglomerated forms on a frontier.

Looking back on what we know about Kofyar pioneer settlement, a central fact is that none of the catalysts for agglomeration were irresistible. There was neither the powerful undertow of central functions nor the necessity of defensive nucleation. There was simply very little impetus for dispersion, because cultivation was extensive and land tenure uncontested; this allowed some of the individual settlements to be prodded toward agglomeration by forces that were, in some cases, relatively weak. This situation changed dramatically over the next few decades.

Evolution of Ungwa

Soon after their arrival, Kofyar migrants on the Namu Plains began to treat the landscape as comprising named sociosettlement entities above the level of the individual compound, called *ungwa*. The ungwa continues to be a basic organizing principle of Kofyar settlement throughout the Benue Lowlands today. Ungwa boundaries are not marked in any way, and from a material standpoint, the ungwa are "invisible villages" (Holzall and Stone 1990).

The Kofyar word *ungwa* is taken from the Hausa *unguwa*, a term with no direct English equivalent. *Unguwa* refers to a settlement grouping, neighborhood, or town ward, implying a certain formality and distinct identity. The Kofyar term *ungwa* refers both to an area on the landscape that is usually (but not always) contiguous and to the households farming within that area. The ungwa is a political, ecological, and social unit.

In 1963 there were six ungwa in the core area (see fig. 10.1). The number of compounds within the ungwa ranged from 16 in Ungwa Long to 40 in Bubuak. Ungwa have proliferated and subdivided since then, producing the pattern visible in figure 6.2 by 1985.

Ungwa partitioning is linked to population growth, but it is often triggered by social friction. Koedoegoer Koegoen (KDG Koegoen) seceded from Bubuak in the 1960s when there was a conflict between the ungwa head (*mengwa*, described below) and his assistant (*maidaiki*); the latter became mengwa of the new ungwa. Because such disputes are often organized along microethnic lines, they serve as a mechanism for increasing microethnic homogeneity within ungwa (chapter 10).

The evolution of the cultural landscape is reflected in the nomenclature. The following survey of place names on the Namu Plains illuminates various aspects of the landscape's history.

Ungwa Goewan is named for the area of Kwa village (or division of the Kwa supravillage) that was home to the builders of the macrocompound called Goewan gari. Ungwa Goewan is sometimes called Rafin Gogo, Hausa for Baboon River, referring to the baboons that plagued early settlers there. I have described the early farmers' desire to have this area generally controlled by Goewan people, and this is by and large the case today (chapter 10 gives microethnic breakdowns of contemporary ungwa).

Ungwa Long is named for Long Kwa (chief of Kwa) Doedel, who spearheaded the migration into the Namu plains in the early 1950s. The area is sometimes called Ungwa Koplong (*kop X* refers to the former residence of X, or the place where X died). It was actually Doedel's brother, Dakoeng, who was mengwa of this first ungwa, as it was decided that Doedel could not be chief and mengwa simultaneously. There are several ungwa named for the first settler there, much as Norse settlements were named for the Landnamsmann, the first settler (McGovern 1994).

Duwe and Koprume were named for Tiv who abandoned the area as the Kofyar were moving in.[6] The Tiv are known for practicing shifting cultivation with shifting settlement, contrasting with the Kofyars'

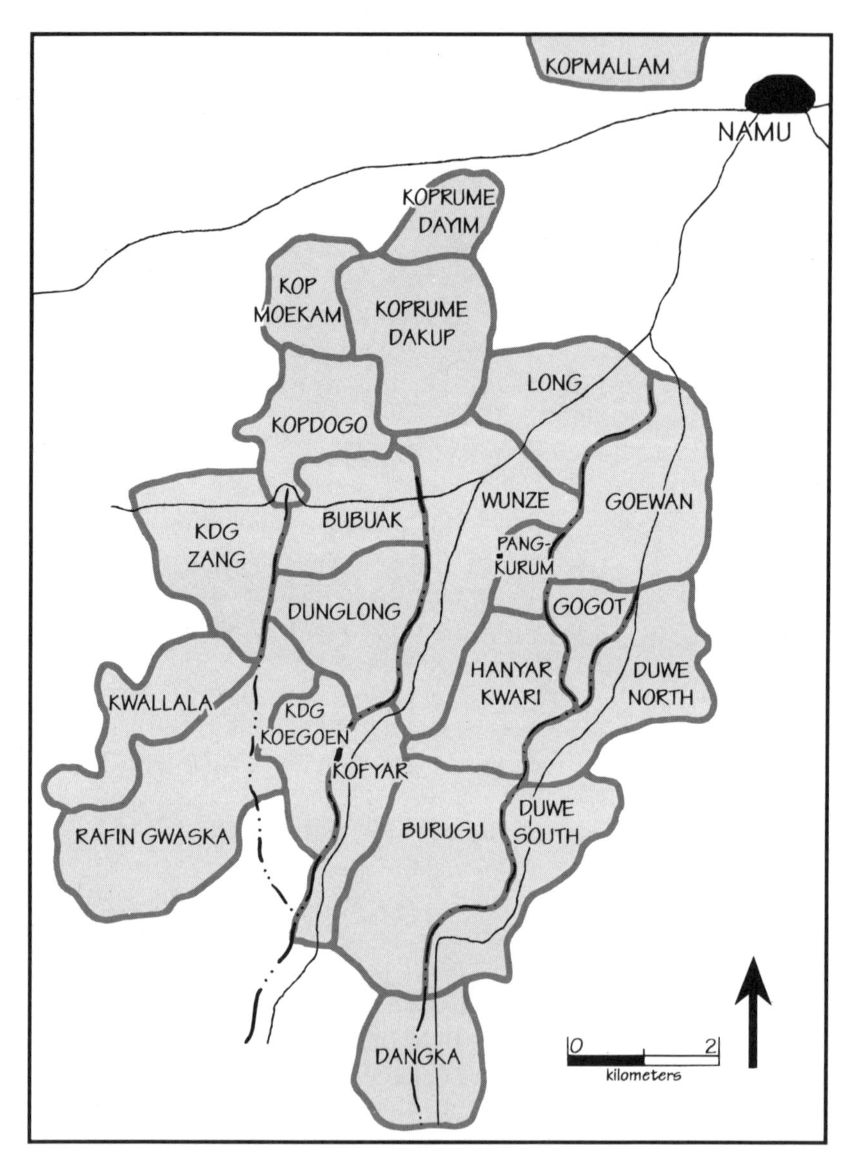

Figure 6.2. Ungwa in the core area, 1985.

proclivity for intensive farming. The presence of enduring Kofyar settlements with intensive cultivation, on spots named for Tiv who presumably abandoned them to avoid intensification, highlights the contrast in the two ecological strategies (Stone 1993a).

Kopdayim, northwest of Namu, is named for Dayim of Doemak, who led settlers into there in the 1940s. He left for the Namu Plains in the early 1950s and was one of the leaders of the Doemak colony that started up in the Koprume area at about the same time the Merniang-dominated colony was starting to the east.

The Kofyar term *goejak* and the Hausa word *dangka* both refer to muddy landscapes. Dangka and most of Goejak were settled relatively late, and their names probably reflect why the Kofyar were unenthusiastic about farming there.

Kopmallam is named for Mallam Musa Gokla, the early Kofyar Muslim who led farmers into this fertile area at about the same time non-Muslim settlers were moving onto the southern plains. The proximity of Kopmallam to Namu reflects the Muslims' strong attraction to the central functions of towns, including the mosque and common Hausa pursuits of crafts and trade. Today there is a sizeable Muslim community in Namu, many of whom still farm near Kopmallam.

Political Organization of the Ungwa

Each ungwa has a headman, or mengwa (from the Hausa *mai ungwa*, or ungwa head). The mengwa is not a chief, and the chiefly title *long* is not used for mengwa. Yet the mengwa was comparable to a homeland chief in that he handled administrative functions such as tax collection and announcements; also like a homeland chief, the mengwa's influence and political clout were quite variable. In the homeland, only a few chiefs managed to garner real power and wealth. The most notable example was Doedel, the paramount chief of the large population to the east of the escarpment, who had 80 wives at the time of his death. His position had been invested with abnormal influence by the colonial government. By contrast, the chief of the large hill village of Bong had a small household, no unusual wealth, and practically no authority beyond that of any elder in the community.

The authority of mengwa on the frontier varied similarly. When asked for permission to conduct our census, some unilaterally authorized our work and instructed all ungwa household heads to cooperate; others called a meeting of the ungwa elders to discuss the matter. In one particularly democratic ungwa, we were formally interviewed by the *daskagam* (male ungwa elders) before permission was granted.

On the early frontier, it was the mengwa's job to collect and deliver tribute from each household to Longjan, the chief of Namu, who was the nominal landowner. After the earliest wave of pioneers, who were granted land directly by Longjan, newcomers would approach the mengwa for land. The mengwa could point out an available plot, but he was not considered to own the land; this is reflected in Kofyar discussions of settlement history, in which they refer to mengwa "showing" them land, rather than giving it.

Some ungwa had an assistant mengwa. This position was sometimes called maidaiki, the name used for the assistant chief in Namu, from the Hausa political hierarchy. These assistants often help with the collection of taxes. For most decisions that affect many people within the ungwa, the decision-making body is the daskagam, normally constituted by the household heads. It is the daskagam that resolve disputes, oversee the scheduling of mar muos, and grant permission for anthropologists to work in their midst.

Among the roles played by the ungwa, one of the most important is in the organization of agricultural production. Discussion of this will have to await descriptions of the agricultural system and the social solutions to its labor demands (chapters 7–10).

7

Land Pressure and Intensification

In the three decades following the establishment of Kofyar pioneer settlements on the Namu Plains, the trickle of migrants turned into something of a flood. At the same time, population grew at the accelerated pace characteristic of frontiers (e.g., Easterlin 1976). From 1972 to 1985, Kofyar population reflected in the Namu District tax lists climbed from 7050 to 12,602, with another 11,364 Kofyar listed for adjoining Kwande District. There were probably more than 30,000 Kofyar farming in the Benue Lowlands in the mid 1980s.

Three decades of cropping have also taken their toll on agricultural productivity, and many areas have seen a steady rise in land pressure. The result has been both settlement spread and agricultural intensification. The initial colony in Ungwa Long has grown into what I call the core area, comprising the numerous ungwa shown in figure 6.2; this area is the main focus of this study. This chapter describes the agricultural system and its connections to the pattern of settlement.

Land Pressure

It is clear that overall population density climbed sharply in Namu District between 1951 and 1985, but measuring the increase in a meaningful way is not as straightforward. The prima facie definition of population density as population per unit area is workable for regional and global comparisons (Lagemann 1977; Pryor 1986). But it is not especially satisfactory for measuring the kinds of land competition that affect farmers' settlement decisions. For instance, untillable land is reflected in measurements of population density but may have little effect on the settlement and agricultural tactics.

In assessing crowding on the Kofyar frontier, we also have to deal with the perpetual flux of population between homeland and frontier. The earliest farms were seasonal outposts, where part of a household would work until the harvest and then return home. But by the 1960s, some of the bush farms were turning into yearlong residences. This was the norm by the mid 1980s, and most hill villages had been depopulated or abandoned. But throughout this history there have been many households maintaining both homeland and frontier farms, usually with part of the household rotating between the two. Any attempt to gauge frontier population must take this flux into account.

Netting's migrant bush farm censuses from 1961 list how many months each household worker spent at the frontier farm. (These censuses do not list children, and indeed children were usually left behind on the home farm.) I have calculated the mean size of the frontier household in 1961 from a random sample of 25 bush censuses. A person was counted as a resident if he/she spent five months or more on the farm and was counted as a half resident if he/she was present between one and four months. In almost all cases the head was counted as a full resident. The sample mean was 3.2, with a range of 1.5–4.0. There are 235 Kofyar compounds in the area shown in figure 6.1, yielding a weighted population estimate of 752. Using figure 6.1, I have calculated a weighted population density of 12/km^2 for the settled area southwest of Namu (G. Stone 1988). The early frontier had what one might call negative land pressure; land was so productive and abundant that concerns of present or future land competition were overwhelmed by the need for more farmers in the labor pool. Pioneers went to some lengths to encourage further migration.

By 1984 the demographic picture had changed considerably. The 1984–85 household census that my coworkers and I conducted contains breakdowns of each individual's economic activities and where they live. I have used this census to weight each individual's presence at each farm, yielding a weighted residential population. I have also computed a weighted farming population, based on the amount of time individuals 14 years of age or older actually spend farming.[1]

Table 7.1 gives 1984 population figures for 15 ungwa in the core area. The best calculation of population density for this area is 101/km^2, more than an eightfold increase from the early 1960s. This is comparable to the population density of the homeland overall in that period, although

Table 7.1 Populations for Core-Area Ungwa

	Total Compounds	Compounds Censused	Avg. WRP[a]	Avg. WFP[b]	km^2	Density/km	
						Residents	Farmers
13-ungwa total[c]	513	477	6.31	3.49	32.11	100.8	55.7
15-ungwa total	563	496	6.40	3.52	36.22	99.5	54.7
Kwallala	29	26	7.4	3.6	2.56	83.4	40.7
Goewan	46	45	6.0	3.3	3.31	83.9	45.8
KDG Koegoen	25	25	5.2	2.9	1.48	88.6	48.9
Dunglong	35	32	6.6	3.8	2.55	91.2	51.8
Bubuak	34	34	5.7	3.7	2.08	92.6	61.1
Long	35	35	7.0	4.1	2.63	93.2	54.0
Duwe North	36	30	6.7	3.7	2.53	95.5	52.6
Wunze	72	69	6.7	3.6	4.80	100.3	53.8
Hanyar Kwari	49	35	4.8	2.8	2.36	100.3	57.4
KDG Zang	44	41	6.7	3.6	2.84	103.9	56.5
Kopdogo	39	37	6.9	3.4	2.17	124.8	61.9
Koprume-Dayim	26	26	6.0	3.2	1.18	132.6	71.2
Kofyar	43	42	5.9	3.4	1.62	155.6	90.6
Gogot	20	6	7.1	3.7	0.93	153.0	80.3
Koprume-Dakup	30	12	9.6	4.8	3.18	90.8	45.2

[a]Weighted residential population (WRP) gives a weighting of one to each full-time resident at the farm and lower weights to part-time residents such as college students and wives who spend part of each year at a farm in the homeland. Children are included.
[b]Weighted farming population (WFP) is based on the amount of time individuals 14 years of age and older spend farming.
[c]The 13-ungwa totals exclude Koprume-Dakup and Gogot for which our samples are less than 50% and possibly biased.

it is higher than that of the hills and lower than that of the plains (Stone et al. 1984:95). An estimated 70% of the land in the core area was being cultivated in 1984.[2]

Land use is probably more intense in the core area than in any other frontier area settled by the Kofyar. By comparison, approximately 50% of the land in Ungwa Mangkogom, south of Kwande, was in crops in 1984.[3] Mangkogom has had a relatively high abandonment rate (chapter 11), and it probably has one of the lowest cultivation ratios of any Kofyar frontier area. Still, even there the *R* value (a measure of cultivated land to fallow land) clearly exceeds the *R* of 33 (i.e., 33% of land in cultivation) that is cited as a significant threshold in tropical savanna farming (Ruthenberg 1980:58; Guyer and Lambin 1993:840).

Crops and Agricultural Tactics

This section provides an overview of the agricultural system in the core area as of 1985, broken down by crop. My references to time investments pertain to average time per person per day; more detailed descriptions and figures appeared in Stone et al. 1990 and M. P. Stone et al. 1995.

Grains

As in the Kofyar homeland, pennisetum millet and sorghum are the major grains. Sorghum is the Kofyar dietary staple, and it is only occasionally sold or made into beer. Millet is made into beer (for consumption or sale), sold, and eaten (in order of importance).

Millet and sorghum are mostly intercropped, although sorghum is sometimes planted alone. Clearing work begins on the grain fields during the dry season, taking up more than an hour daily during the weeks preceding the rains. As soon as the rains arrive (near the end of March in 1985), everyone works long hours planting the alternating sorghum and millet seeds. Pennisetum millet has a short (four-month) growing season, and the crop must be in as soon as the rains begin and out by August when the rains slacken. Sorghum is photoperiodic, flowering more or less independently of planting date (Kowal and Kassam 1978:239), and it may be put into unused strips and patches around the farm even as late as July.

The work schedule in the month of May is dominated by the making of parallel ridges through the grain fields. There are different ways of accomplishing this; full rows may be made prior to planting (in an operation called *pian*), or dirt from between the plant rows may be hoed between and around the young plants (*ogot*). The field is usually weeded at least once before August, when the millet is rapidly harvested, dried, and stored in specialized millet storage structures (*lu maar* or *paar maar*). The field is weeded at least once more before the sorghum harvest in December.

Maize is grown in smaller quantities, often intercropped with yams.

Rice has been grown as a cash crop since the early days on the frontier, and it was grown by a quarter of the households censused for this study. It is planted in streambeds, either in standing water or on moist earth. Field preparation ranges from simply turning over the topsoil to making widely spaced waffle ridges. The seed is broadcast, usually while the field is being prepared. This usually takes place in July. The fields are weeded once or twice in September or October, and the harvest takes place in November or early December.

Yams

Yams are a favored staple throughout Nigeria, and the market for yams was one of the early attractions for the Kofyar on the frontier. This crop, not grown in the homeland, now absorbs 36% of all agricultural labor. On many farms, a large portion of the land near the compound is alternately planted in yams and sorghum-millet.

Yam field preparation requires making knee-high earthen heaps (*bunga*), one of the most arduous tasks in the calendar. Bunga are nearly impossible to make when the soils are dry and hard, and so farmers may wait for the onset of the rains to build and plant the heaps. But a better strategy is to make the bunga at the tail end of the rains—September or October—of the year before the yams are to be grown, constructing them right in the grain field that will then rotate back to yams (fig. 7.1). Thus, next year's heaps are made between and around this year's sorghum plants; the heaps are then planted in the dry season. This ingenious strategy circumvents a disastrous early-season labor bottleneck; it allows the heaping operation to serve as a weeding of the sorghum; and it helps to buttress the sorghum stalks against winds. For these reasons, the September-October bunga work is very desirable. Almost all of it goes toward household heads' fields, and women who are growing their own yams must usually wait until the next year's rains (M. P. Stone et al. 1995). Yam fields are usually weeded twice during the growing season. The harvest lasts through the dry season.

Figure 7.1. Intensive farming in Ungwa Kofyar. In September, when this photograph was taken, the field reflects the intricate crop scheduling. Millet has been harvested, and the stalks have been left in the field to be used in capping the heaps (bunga) recently made for yams, which will be planted in the dry season. Sorghum is still growing and will be harvested in December. A quick crop of sesame is being squeezed in by planting in the newly made heaps. Peanuts, grown as an understory crop, have also been harvested.

Minor Crops

The Kofyar grow a wide variety of minor crops that fill in gaps in their fields and work schedules. As farmers spread into new areas, and as intensification alters the agricultural calendar, the farmers maintain a keen interest in experimenting with new subsidiary crops.[4] Included in the roster of minor crops are root crops such as sweet potato, cocoyam, and manioc; tree crops such as locust bean, mango, papaya, and banana; and legumes such as cowpea, peanut, bambara nut, and green beans.

Agricultural Workload

How intensive is this agricultural system? Agricultural intensity is often measured by the presence of specific practices (e.g., Brown and Podolefsky 1976), and it is clear that many of the practices characteristic of intensive farming are present. Fields are carefully prepared, and all crops are weeded; increasingly, farmers are using dung/compost or chemicals to boost soil fertility. But the question can be answered more directly with data from the year-long study of agricultural labor (described more fully in the appendix) by my coworkers and I, which provided detailed records of labor inputs for 11 households from the core-area ungwa of Kwallala and Kofyar and an additional 4 households from Dadin Kowa. The labor study not only reveals how much effort is being invested, but gives a detailed picture of how cultivation has been intensified.

Because the crux of agricultural intensification is the replacement of land with labor (or capital) inputs, I begin here by looking directly at labor inputs and mobilization across the Kofyar agricultural calendar. Figure 7.2 shows the average daily agricultural labor input across the agricultural calendar (see Stone et al. 1990 for breakdowns of labor input by crop and task). The total yearly labor input as measured by this study is 1501 hours per adult, which is quite high compared with that of paleotechnic farmers worldwide.[5] Cleave's (1974) survey of African farming shows yearly inputs of 480 to 960 hours. The Kofyar figures fall into the range of intensive wet-rice farmers of Japan (Clark and Haswell 1967).

The labor figure works out to 4.8 hours per day in a 6-day workweek. The actual workday is longer because the figures omit time spent on tasks under 45 minutes in duration, as well as tasks that are not directly agricultural, such as tool maintenance, house construction, and travel between farms.

As intensification has increased, the Kofyar have developed strategies for allowing high labor investments in a packed work schedule while mitigating bottlenecks (Stone et al. 1990). One has been to shift as much work as possible into the dry season. Much of the field clearing, yam planting, and yam harvest occurs between January and March, and figure 7.1 shows that the schedule does not have the sharp

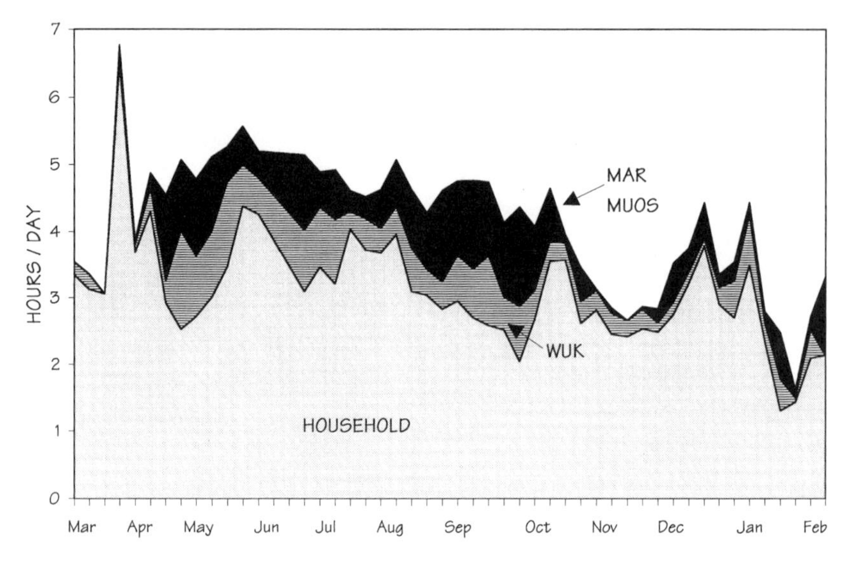

Figure 7.2. Social organization of agricultural labor inputs for all crops, 1984–85.

dry season drop-off in work found where farming is less intensive (e.g., Tripp 1982).

Another important strategy has been to fill in relatively slack time with work on temporally flexible crops. The labor-sample farmers devoted 21.1% of their time to crops other than sorghum, millet, and yams. Much of this time went to peanuts, a predominantly woman's crop that is especially flexible in the timing of its planting, weeding, and harvest (Stone et al. 1990:fig. 6; M. P. Stone 1988a; M. P. Stone et al. 1995).

Methods of intensifying agriculture may share the characteristic of concentrating resources in time and space, but different intensive tactics can have quite different effects on spatial organization (for instance, irrigation and fallow shortening may have opposite effects on settlement). With the Kofyar, intensification has not meant change in technology (which is still based on the hoe, although chemical inputs are being adopted on a small scale) or in the nature of tasks (although weeding has necessarily risen) as much as it has meant change in the density and arrangement of crops. Crops are now packed more tightly in time and space.

This is well illustrated by a single field in Ungwa Kofyar that had been planted in millet and sorghum early in the season. After the labor crunch caused by grain field ridging in May, a woman put in an understory crop of peanuts. The millet was harvested in August and the peanuts in September. In October, heaps (bunga) were put into the field for the next year's yams. The bunga operation not only weeded the sorghum but also left a field of fresh, plantable surfaces, which the owner immediately planted in a catch crop of sesame. Work on five crops has thus been arranged into five separate "shifts" (fig. 7.1).

As these strategies have caused agricultural labor demands to become more densely and complexly packed in time, it has placed an intricate set of demands on the social mechanisms for mobilizing labor. Just as increased cropping affects the relationship between the residence and the plot, evolving social relations of production affect relationships between residences, as described in the next section.

Organization of Agricultural Labor

Let us look at the frontier Kofyars' social mechanisms for providing labor, and at the Kofyars' reasons for mobilizing labor in these ways. I focus here on the three principal labor mobilization strategies that account for more than 98% of all farm work.

Household labor, in which family households work on their own fields, accounts for the bulk of Kofyar agricultural hours. Household fields include both those plots controlled by the household head and those held in usufruct by others in the household. Household labor is applied by individuals or groups usually numbering five or less. This type of labor is easily mobilized and highly flexible.

Mar muos, based on the homeland labor parties of the same name, are large neighborhood work groups, also known as festive labor parties (Erasmus 1956; Saul 1983). These gatherings typically involve 30–60 workers but may exceed 100. They are characterized by a spirit of friendly competition and an almost frenzied pace of work. All present are served millet beer after the work (fig. 7.3). Mar muos usually must be scheduled weeks in advance, and they require several days of brewing work by the women in the household.

Wuk exchange-labor groups, typically ranging from 5 to 20 in size, are between the more formal mar muos and small-scale, flexible household labor. Wuk are membership groups or voluntary associations whose participants take turns working on each other's fields. The workers are repaid with reciprocal labor (which is carefully noted) at later

Figure 7.3. Dancing after a large mar muos in KDG Zang. The household that hosts a large work party, especially for an arduous task such as bunga (making yam heaps), usually hosts the festivities afterward. The host is obligated to provide millet beer for all, and there is often music and dancing, and occasionally the slaughtering of a small pig.

meetings of the group. Most households belong to a wuk with their neighbors, sending various household members to each labor event. Individuals also form wuk groups, usually along age and sex lines, that meet to work on individual plots.[6]

The labor budget recorded in the study sample comprised 74.5% household, 11.5% wuk, and 14.0% mar muos. Group labor is especially important during the rainy season; exactly one-third of the farm labor between mid April and mid October is mobilized by suprahousehold groups (fig. 7.1). But why is labor pooled in the first place?

Net Labor

There is first the fundamental question of whether labor exchange has an economic basis at all: where is the advantage in putting in *n* days on neighbors' farms so that *n* neighbors will put in a day on mine? In fact, the Kofyar would seem to suffer a net loss in field labor because of the time women have to spend on brewing beer for the group (M. P. Stone et al. 1995).

Part of the Kofyars' answer is that people work harder in work groups, especially in the festive atmosphere of the mar muos. There are ample inducements for hard work at the mar muos. The work may be preceded by the spectacle of a few young men making their way across the field, each hoeing a line of yam heaps at a frantic pace and occasionally stopping to shout energetically; this provides a model of industrious field work and also divides the field into sections for the other workers, who compete in teams. Sponsors sometimes hire drummers to encourage the workers to keep a fast pace (see fig. 3.3). The least subtle technique that I saw for boosting labor productivity was the masquerade character who attended a chief's mar muos for millet storing; the character carried a sorghum stalk, and part of his role was to whip anyone whose work pace was too slow.

A related issue is the quality of work, which is relatively high with the Kofyar (Netting et al. 1989:308). Complaints about the quality of work by festive groups are usually based on comparison with work done by households on their own fields (Erasmus 1956:456), or they concern extensive plots that do not require fine techniques (e.g., Saul 1983). For the most part, the Kofyar show a strong preference for festive or exchange labor over hired labor, which is expensive and requires constant monitoring. This is especially true of complex intercropping operations (such as making yam heaps in the sorghum field) that require particular skill and judgement, or the millet storing (described below), which would never be done with hired labor.

Simultaneous Labor Demands

Intensification theory stresses total labor demands, but in some ways the arrangement of labor demands through time is more important. In chapter 3 I discussed how farm operations that can be accomplished by low labor inputs over an extended time are said to have linear labor demands. Operations that require a large number of workers at once have simultaneous labor demands; this includes operations that simply require that a lot of work is accomplished in a short time (simple simultaneous), and those that require that different tasks are completed at the same time (complex simultaneous; see Wilk and Netting 1984; Jochim 1991). Barn raising and field burning are well-known examples of simultaneous labor demands that are usually handled with group labor.

The Kofyars' use of group and household work details to meet differing labor demands is seen most clearly in the harvesting and storing of millet. Millet is harvested in August. The work of cutting the grain

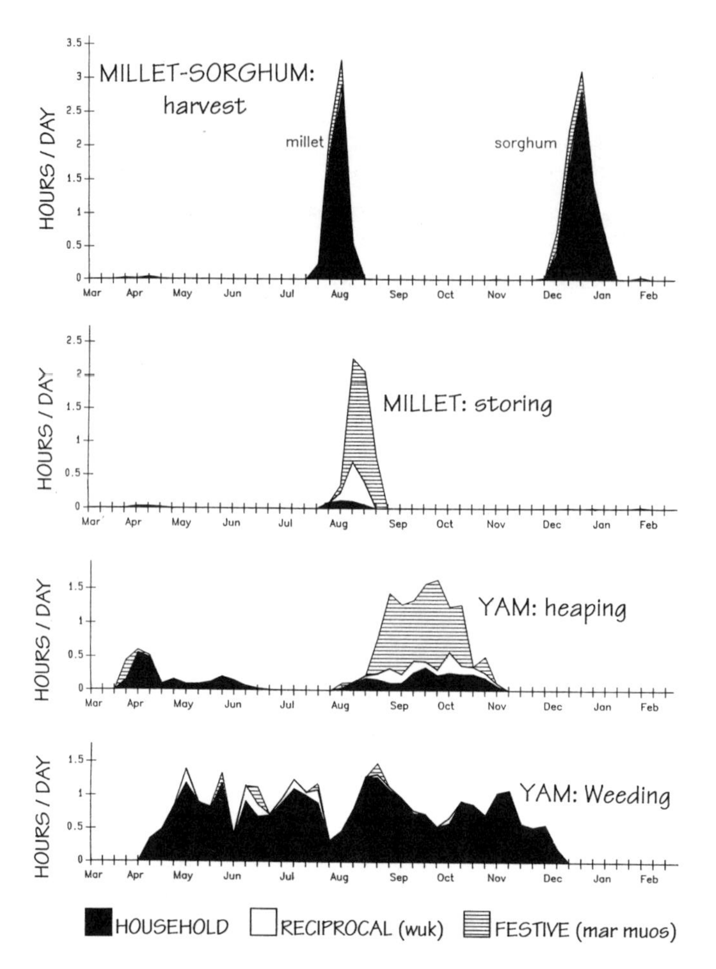

Figure 7.4. Social organization of agricultural labor inputs for selected crops, 1984–85.

heads, tying them into bundles, and carrying them back to the compound can be done piecemeal, and the job is done almost entirely by household labor, not larger work details (fig. 7.2). But the harvest is immediately followed by the *lang maar* storage operation, with its complex simultaneous labor demands. Grasses are gathered for ropes and thatching; ropes are braided to tie bundles of seed heads while thatch mats are woven; the bundles are heaved up to workers atop the small circular millet house (*lu maar*) to be arranged in a high conical pile. When the pile is complete, the thatching is secured around it and a small fire is

Figure 7.5. A large mar muos in Dadin Kowa for millet storing (lang maar).

started on the dirt floor of the millet house to begin drying the crop. Because most of these operations must be conducted simultaneously, mar muos are particularly adapted to this task and account for more than 72% of the work (fig. 7.4). Millet storage parties can put away millet for several households in one day, often beginning early in the morning and working into the late afternoon (fig. 7.5).

The only time a Kofyar asked me for help on his farm was when I stopped by to talk with a man who was alone in his courtyard, tying millet into bundles. This man was a member of the Protestant church, which forbade production and consumption of the muos (millet beer) that is the lifeblood of festive labor arrangements. Group labor has been greatly reduced (but not eliminated) among the Protestants, and he was faced with the problem of both throwing and catching the bundles—the epitome of complex simultaneous labor demands.

Labor Banking

The mar muos system allows labor invested into millet cultivation to be withdrawn in the form of group labor. Millet is a particularly advantageous crop to grow; interplanting it with sorghum increases returns per unit of both labor and land, without a proportionate rise in labor demands during bottlenecks (Stone et al. 1990; Richards 1985; Norman et al. 1982). Its value is enhanced by its flexibility. It can be eaten, and is available long before other staples are harvested. It can be marketed, and is the Kofyar's second major cash crop after yams. It can also brewed into beer for sale, mar muos, or social occasions. "Banking" labor as millet allows the farmer to mitigate fluctuations in the availability of household labor. The Kofyars' intensive cultivation strategies are planned tightly enough that the loss of one adult to illness during a key part of the season may endanger an entire crop.

We can see, then, that the premium on dispersion of the population into small sociosettlement units is paralleled by several advantages of pooling labor across those units. Many of these advantages were described by Erasmus long ago, although he saw labor pooling as a disappearing phenomenon (Erasmus 1956). I would be surprised to see it disappear from the Kofyar production system anytime soon.[7]

Degree and Variation in Intensity

A closer look at farming in the core area reveals variations stemming from differences in population density. Let us compare Ungwa Kofyar, which has the highest population density of any core-area ungwa, with the Kwallala section of Goejak, which has the lowest (table 7.1).

Ungwa Kofyar is an early-settled area, some farms having already been claimed by the late 1950s and some residences having been started by the early 1960s. It runs along a watershed between two converging streams, giving it a linear shape. This 1.6 km^2 area had 43 farmsteads at the time of the study, and its population density of 155/km^2 was one of the highest on the frontier.

Kwallala, by contrast, is a relatively late-settled area because of its limited access to water. Most of the population in Kwallala did not arrive until the 1970s, and many of its farms still contain uncut woodland. Kwallala has 32 households on 2.6 km^2, and its population density of 86/km^2 is one of the lowest on the frontier. Although Kwallala is a discrete area, it constitutes, along with KDG Koegoen, the single noncontiguous ungwa of Goejak. Because of the history of settlement, Kwallala and KDG Koegoen recognize a single mengwa and often collaborate in group labor parties.

Table 7.2 compares agricultural inputs by crop for Kofyar and Kwallala. Approximately 80 percent of the total labor occurred on the farmer's home farm, an average of almost exactly one task on the home fields per day. In fact, farmers often conduct several operations per day in the rainy season and sometimes do no farmwork in dry-season days; nevertheless, the general pattern is definitely one of frequent and regular travel to the home plot.

The longer-settled Kofyar has undergone farm fragmentation,[8] reducing the mean farm size to an 4.75 ha as compared to 6.31 ha in Kwallala. By African standards, cultivation is intensive in both areas, but it is demonstrably more so in the more crowded ungwa, where returns on labor are lower. Ungwa Kofyar farmers are putting 21.4% more work into their own farms and 6.3% more work into neighbors' farms, for an 18.2% greater total labor outlay on farms that are only 75% as large as the average Kwallala farm. Labor bouts in Ungwa Kofyar are comparatively short, and marginal returns to labor are decidedly lower. Table 7.2 shows that the average hour in the fields in Kwallala produces 81% more yams, 4% more sorghum, 55% more millet, and 40% more rice than in Kofyar.[9]

The Ungwa and Agricultural Production

The ungwa divisions play an instrumental role in the mobilization of communal labor. Out of 2616 recorded bouts of work on other farms, 94% occurred on farms within the worker's own ungwa. The pattern holds for farmsteads located near ungwa edges as well as ungwa centers, confirming that ungwa are agricultural labor pools. Because this mobilization of suprahousehold labor normally occurs within ungwa boundaries, one might expect the size of the ungwa to reflect the size of labor parties; it does, although this requires some explanation.

Based on farmers' own estimates, the median mar muos size is 50 workers.[10] This is much closer to the median of 35 households per ungwa than to the median adult population of 238 per ungwa and reflects the rules of participation in mar muos. Because the household's commitment to group-labor mobilization must not jeopardize the operation of the household farm, no household is required to send its entire adult population to any mar muos; except for special occasions, one may pass up a mar muos in favor of pressing chores at home.[11] Neither can any household withdraw from the communal labor pool. A household's repeated absence gives the others something to grouse about over beer after the work, and those who repeatedly skip mar muos may be fined (generally payable in millet beer) or even ostracized by the community. This ensures reliability in the communal labor groups without

Table 7.2 Labor, Movement, and Production in Kofyar and Kwallala in Labor-Sample Households

	Kofyar	Kwallala
Population per km^2	154.9	85.6
Mean Farm Size (ha)[a]	4.75	6.31
Households in Sample	7	4
Adults per Household	3.6	4.3
Agricultural Production[b]		
Household Averages		
Yams (individual)	2571	5500
Sorghum (bundles)	44.1	55.0
Millet (bundles)	29.5	50.8
Rice (bags)	2.7	4.0
Per Capita (adult) Averages		
Yams	833	1275
Sorghum	14.1	12.5
Millet	8.7	11.4
Rice	0.87	0.94
Per 1000 Person-Hours[c]		
Yams	524	948
Sorghum	8.9	9.3
Millet	5.5	8.5
Rice	0.5	0.7
Per Hectare		
Yams	529	471
Sorghum	4.2	4.1
Millet	5.2	4.7
Rice	0.5	0.8
Hours (per year)		
On Own Farm	1288	1061
On Other Farms	302	284
Total	1590	1345
Trips per Person (per year)		
To Own Farm	391	324
To Other Farms	59	59
Total	450	383

[a]The average farm sizes of 4.2 and 8.8 ha for Ungwa Kofyar and Kwallala given in Stone 1991 (p. 347) are ungwa-wide averages. The figures in this table are the average size of the labor-sample household farms (n = 7,4 for Ungwa Kofyar and Kwallala, respectively), which give more meaningful yield/ha ratios.

endangering the household farming enterprise. The household contribution to normal mar muos averages out to between 1 and 2 workers, which explains the relationship between the size of communal labor groups and the number of households within the ungwa, rather than individuals. The ungwa does operate as a labor pool, but distribution of ungwa sizes reflects a solution to the tension between communal and household labor.

This case shows how communal labor mobilization promotes the existence of, and shapes the size of, formal settlement entities. Yet it also shows how, when the agricultural system favors residences on plots, these formal settlement entities may form within dispersed patterns. Communal labor mobilized in this way can play a central role in food production; there may be an important reliance on communal labor, administered on a relatively formal manner, without any localization of the administration.

---

[b]Figures are sample means, based on data for 1984.

[c]The figures on production per 1000 person-hours were derived by dividing per capita production by the average total labor input. Figures are available for time worked by labor-sample households on particular crops, but using these figures would miss the labor exchange that is important in production. This means that the figures on production per unit time can be used to compare Kofyar and Kwallala but not to compare the farmer's marginal returns to labor on various crops.

8

Intensification, Dispersal, and Agglomeration

Although the conditions of the early frontier placed a relatively low premium on dispersal, allowing at least ephemeral aggregations to be prompted by the various causes I have described, the increasing attraction of the residence to the plot had by the mid 1980s produced a highly dispersed pattern, with the overwhelming majority of the population in small compounds situated near the center of the household farm. This chapter describes and explains the characteristic dispersed pattern, but it begins by examining the exceptions, in which settlement is not dispersed. At the end of the chapter I will return to the issue of agglomerated settlement, especially the distinctive settlement form the Kofyar call the "company."

Nondispersed Settlement

One exception to the pattern of dispersal is scattered cases of temporary usufruct rights granted to people residing elsewhere. Some women in land-poor households ask friends or neighbors to loan them space for planting. Men from land-poor households, or part-time farmers, may do the same. When land is loaned it is generally with the stipulation that no buildings be erected, as this may be taken to show establishment of long-term rights to the plot (Netting et al. 1989).

Although farmers cultivating others' land are still, in a strict sense, living in dispersed settlements, their pattern is inconsistent with the spirit of my definition of dispersal as living close to one's own cultivated fields. I do not have measurements of what proportion of the land in the core area is worked by nonresident farmers, but I suspect it to be less than 5%.

A second exception is town dwellers who cultivate land in the countryside. Kofyar living in Namu travel to plots in areas such as Kopmallam, typically between 1 and 2 km from their town residence. There are also a few plots in the core area farmed by town dwellers. Namu-dwelling Kofyar are predominantly (69%) Muslim, which is recognized as a change not just in religion but in ethnicity; the Kofyar term *bakwa* means both Hausa and Muslim. Bakwa Kofyar usually adopt Hausa or Arabic names and speak Hausa. They also reduce the role of farming within the household economy and move into towns. Almost half of the Kofyar adults living in Namu (109 out of 246) engage in no farming at all, and less than one-third (73 out of 246) are full-time farmers. For this group, the premium on dispersed settlement is much lower than for the core-area farmers, three-quarters of whom are full-time farmers (1730 out of 2321); only 58 do not farm. The pull to dispersion is further reduced by the availability of transportation: 59.4% of those who farm out of Namu (54.3% of the full-time farmers from Namu) have a bicycle or motorized transport. It is also likely that Namu farmers are cultivating their plots less intensively than are the dispersed settlers in the hinterland.

In sum, the incentive for on-farm residence is lowered by involvement in the nonagricultural economy, by mechanized transport, and probably by less intensive land use. The pull toward dispersion is mild enough that, for an estimated 5% of the Kofyar, it is overridden by the cultural and economic pull of the Muslim community in Namu (see Bruce 1982 on the economic strategies afforded by conversion to Islam).

The third form of nondispersion is the multiple-household groupings that the Kofyar call "companies," which are in some ways parallel to the macrocompounds of the early frontier. This settlement phenomenon warrants a closer look, and I return to it at the end of the chapter.

Dispersion

The rule to which I have described the exceptions is that settlements in the core area are dispersed, with the individual household residing near the center of a contiguous farm, which absorbs most of the household's agricultural labor. These farms are not large enough that we should ex-

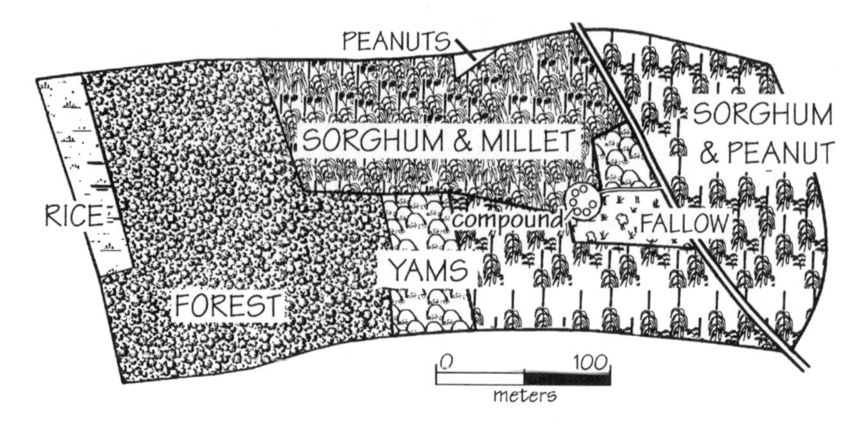

Figure 8.1. Example of frontier farm layout. This farm is located in Kwallala, one of the few ungwa with remaining forest in 1985.

pect pronounced concentric zonation of land use, yet land-use intensity does decline with distance from the compound. Where there is still unfarmed bush, it is invariably on the fringes of the farm, whereas the chief crops of yam and millet/sorghum are usually grown relatively close to the residence (fig. 8.1). This means that settlement is not only dispersed in that individual units of production are separated onto their own farmlands, but that population tends to be located closest to the spots most intensively cultivated.

Settlement Gravitation and Farm Shape

Chapter 3 distinguished between settlement dispersion and gravitation. Attraction of settlements to agricultural plots necessarily pulls people away from each other, militating against settlement gravitation. In some models this should pose no problem; indeed, settlement repulsion is expected for hamlets in Flannery's "rules" of formative Oaxacan settlement and for some farmsteads in Hudson's theory of settlement evolution (chapter 2).

Social physics models (Crumley 1979) expect a relationship between intersite distance and intersite interaction, and in the Kofyar case, site spacing is indeed influenced by interaction between sites. However, it is not a generic phenomenon of interaction but specifically agricultural collaboration that produces the gravitation, which in turn strongly affects agrarian settlement patterns. Consistent with my approach to agrarian settlements in the context of agricultural production, I will

Table 8.1 Size and Population of Core-Area Farmsteads

	n	Mean	Standard Deviation
Farm Size	14 farms	5.0–5.5 ha[a]	1.75
Walled Structures	771 compounds	5.2	3.7
Total Rooms	771 compounds	5.4	3.4
Children (<14)	578 farms	2.8	2.4
Women (14–65)	578 farms	2.1	1.4
Men (14–65)	578 farms	1.3	0.8
Adults (>65)	578 farms	0.1	0.3
Total Household	578 farms	6.2	3.7

[a]An opportunistic sample of 10 farms in Ungwa Kofyar and 4 farms in Kwallala yields a mean farm size of 5.0 ha. Ungwa Kofyar has the highest population density in the core area and an atypically high percentage of fragmented farms; the average farm size in the entire core area must be somewhat larger. My analysis of aerial photos yields an estimate of 5.5 ha per farm in the core area, which I regard as a solid estimate.

look not just at the spacing of Kofyar compounds, but at the shape and arrangement of Kofyar farmsteads.

Table 7.2 shows that the Kofyar farmer invests well over 300 separate labor bouts on his or her own farm annually; this labor investment on a nonfragmented farm creates a high premium on low residence-to-plot distance, and the overwhelming majority of settlements are dispersed. Table 7.2 also shows that core-area farmers are making 59 trips to work on other farms each year—well over a trip per week during the main farming season, with a few crucial weeks averaging more than 3 trips. Overall, these farmers are putting in nearly one hour on a neighbor's farm for every four hours on their own.

The effect of this "agricultural movement" should be consistent with the principle underlying both central place theory and von Thünen's land-use theory: frequency of access promotes proximity, in this case causing residences to be attracted primarily to their plots but secondarily to each other. Let us consider what would be the optimal spatial solution to these competing attractions.

In response to the pull to one's own farm, the optimal farm shape on an imaginary idealized plain would be circular, as von Thünen showed. But whereas von Thünen's concern was with an individual settlement—"the isolated state"—my concern is with settlement pattern, and here I consider only polyhedral shapes because farms generally

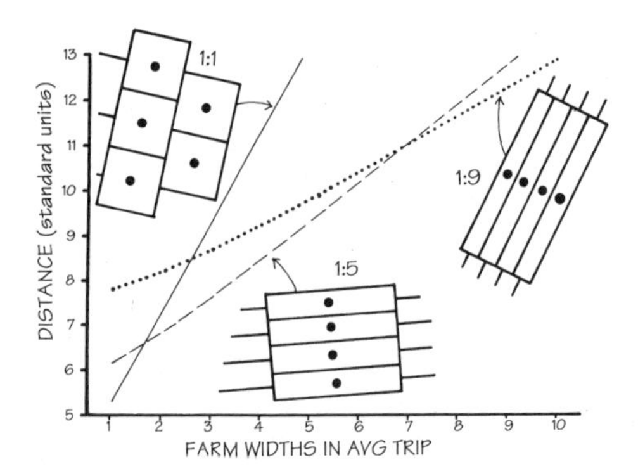

Figure 8.2. Length-to-width ratio and travel costs.

have contiguous borders. The optimal solution to the problem of intrafarm agricultural movement, assuming the geographer's isotropic plain, is a regular polyhedron; the more sides, the more efficient. The optimal solution to both intra- and interfarm movement on the isotropic plain is the hexagonal lattice, for reasons demonstrated by Christaller (1966).[1] But the Namu Plains were not isotropic to Kofyar settlers. As I explain later, settlements were attracted to strongly linear features on the landscape: streams in the early phases of settlement, and roads throughout the evolution of the settlement pattern. Settlement proceeded not like a fan but in strips (fig. 6.1). What is the optimal solution to intra- and interfarm movement in a linear settlement pattern?

To model this, picture rectangular settlements with the same areas (defined as 100 square units) but with varying length-to-width ratios (fig. 8.2). Travel cost (reckoned as the distance a farmer must traverse for a given number of agricultural moves) varies with the length-to-width ratio of the rectangle and with the nature of off-farm travel. For the Kofyar, 20% of all trips are to neighbors' farms. At this percentage, the 1:1 ratio of the square is optimal for intrafarm movement, but travel cost rises so sharply that the square has the highest travel cost if the average trip is to three farm widths away.[2] Optimal farm shape varies with the size of the area in which farmers will move for agricultural collaboration. For farmers whose average trip is three farms away, the most efficient farm shape is 1:3; for five farms away, 1:5; for seven farms away, 1:7; and for nine farms away, 1:9.

Figure 8.3 shows the outlines of the 11 farms in the labor sample for this study.[3] The average width is 109 m (s.d.=31), against an average length of 518 m (s.d.=151). The length-to-width ratios are also shown

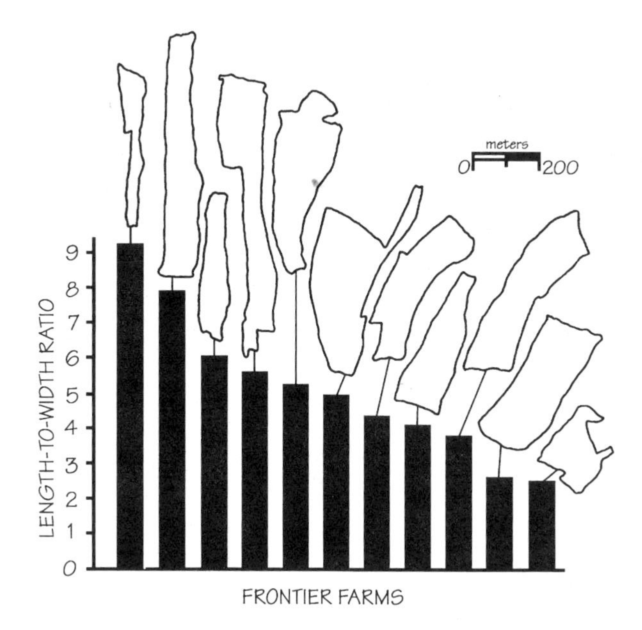

Figure 8.3. Farm shapes and length-to-width ratios for farms of the labor study households in Ungwa Kofyar and Kwallala.

in figure 8.3; the average ratio is 5.1 (s.d.=2.0), and the ratios of 8 of the 11 farms are between 3.8 and 6.1.[4] The length-to-width ratio of 5 is optimal for farmers who expend 20% of their labor on farms that are 4–6 farm intervals away. Given that Kofyar farms average 109 m in width, this means that their shape is optimal for the farmers' interfarm trips to be to destinations 436–654 m away.

What is especially interesting about this finding is that, as I show in chapter 9, Kofyar agricultural movement shows a threshold at 700 m; within this range, there is no decrease in agricultural trips, but beyond it trips decline steadily with increasing distance. On the basis of this limited sample, there clearly appears to be a adjustment between agricultural movement and farm shape.

Settlement Gravitation and Site Spacing

If Kofyar settlement spacing were as neat as the idealized pattern calculated on the basis of figure 8.2, all nearest-neighbor distances would be 124 m. In reality the details of settlement spacing are sensitive to a variety of minor factors. In the first place, because compounds are con-

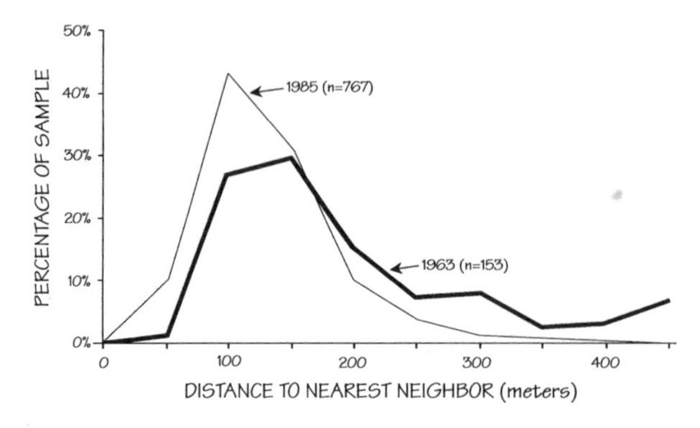

Figure 8.4. Nearest-neighbor distances in the core area in 1963 and 1985.

stantly being founded and abandoned, a snapshot of the settlement pattern at any one time may record what appear to be relatively isolated compounds; this may happen when one family moves into a new area ahead of others who plan on joining them. Even without changing farms, the compound location may be shifted to produce a fresh courtyard surface, to remain close to the part of the farm being cultivated, or to adjust for changes in the household's developmental cycle. Moreover, in some areas farm boundaries extend from the road or path back in one direction rather than spanning the road like the mapped farms; when this happens, there may be compounds across the road from each other.[5] Compound locations are also affected by the fragmentation process; the owner's decision on the size and location of the portion of the farm being sold or given away obviously constrains the location of the new compound.

Still, nearest-neighbor distances in 1985 fall mostly into the vicinity of 125 m (fig. 8.4); 74% are in the 50–150 m range, and 94% are within 200 m. In a sample of 767 compounds, the median is 97 m and the mean 105 m (s.d.=52). This is a very peaked distribution, reflecting the filling in of the settlement pattern with a clear pattern in nearest-neighbor distances.

In contrast, early frontier settlements were less uniformly spaced, and relatively isolated settlements were more common. In a sample of 153 compounds, the mean and median distances are 178 m and 130 m (s.d.=138), and only 73% are within 200 m. The main reason is that the minor agglomeration described for the early frontier was still occurring in 1963.

For instance, a group of farmers from the hill village of Pangkurum met and decided to come to the frontier as a group. They were led by Doebang Bagalta, and most or all of them at first lived together in what is now Doebang's compound. They claimed a contiguous set of parcels near the river, which now form Ungwa Pangkurum (often considered a subset of Wunze). Over the years, the households have dispersed onto their plots in much the same way as did the founders of Goewan gari.

"Companies"

The pull toward dispersion was weak on the early frontier, allowing small agglomerations to develop in response to relatively weak forces (chapter 6). There are contemporary situations where declining population has also reduced the dispersive effect of intensive agriculture to where other considerations have been able to incite the creation of small agglomerations. This has occurred on the frontier and in the homeland as well; the main social force prompting the clustering of formerly dispersed settlers has been religion.

As the Kofyar diaspora has populated the Namu frontier, it has drained population from the homeland villages. The village of Bong, high in the hills of the Jos Plateau, is where Netting lived during much of his fieldwork with the Kofyar in the 1960s (Netting 1968). Although Bong is characterized by a somewhat greater reliance on outfield farming and on the cultivation of maize (because of higher elevation and more rainfall) than other Kofyar "villages," the basic Kofyar homeland pattern of individual farmsteads with intensively cultivated infields is very much in evidence there. Bong is relatively large for a homeland village; in 1966, its 54 households had an adult population of 165.

Bong is one of several hill villages with its own primary school, which acts as a settlement anchor, but it has nevertheless suffered a sharp decline in adult population. The 1984 census for this study shows 48 households with an adult population of 103; the adult population has dropped 37.6%. This brings the modern population density down to the range of the early frontier.[6]

In the late 1970s a new settlement form appeared on the Bong landscape. Six households abandoned their farmsteads and erected a macrocompound, "the Company." Figure 8.5 compares the Company with the ground plan of a traditional Bong compound. Although it is not a large settlement concentration, the Company is still quite notable on a landscape where other settlement is relentlessly dispersed.

Company residents cite religion as the cause behind the Company. A

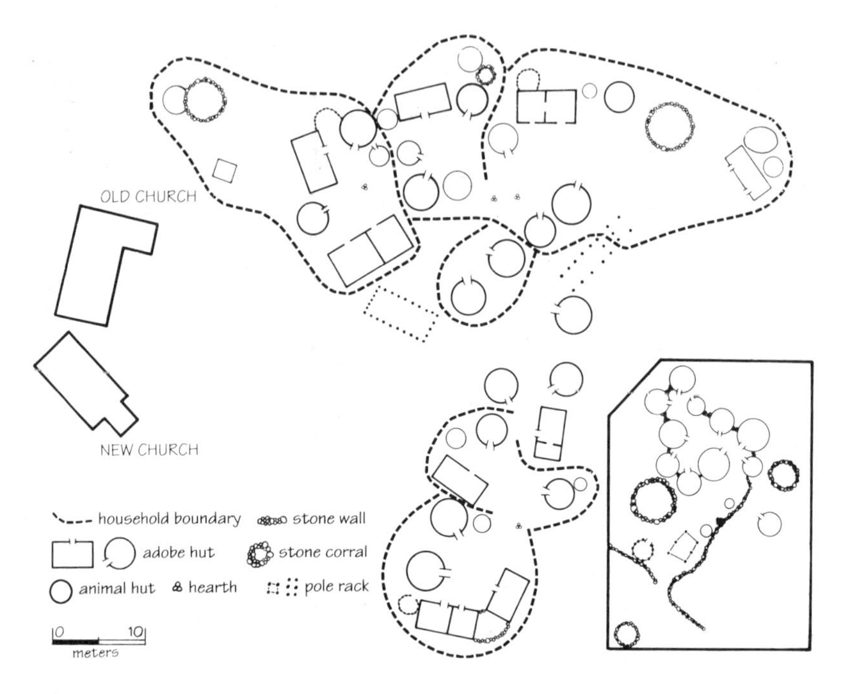

Figure 8.5 The Company macrocompound in Bong village, 1984. The inset shows a typical Bong compound at the same scale.

small faction of Protestants in the community grew through the 1960s and 1970s, eventually leaving their dispersed compounds to form a new social and settlement entity. Members of the Company explained that the most important single factor in the coalescing of compounds was the need to appear in church every morning for prayers. Morning prayers were held before sunrise to keep from interfering with the workday, and walking the rugged terrain in the dark was time-consuming and hazardous. The Company was also close to the primary school, which had been established in Bong partly because of lobbying by Company members.

The Protestants were also increasingly isolated by their proscription on beer drinking. Before Protestantism arrived, millet beer had featured prominently in Kofyar social life, and it had been ingrained in the social organization of production (chapter 7). Unlike the Irish Catholic missionaries, who did not object to Kofyar drinking, the Protestants were strict on the subject of alcohol. This was one factor that checked the spread of Protestantism in most areas; a sample of 4195 adults in 1984 identified themselves as 66% Catholic, 23% traditionalist, 7% Protestant, and 5% Muslim.[7] Bong village, relatively isolated and subjected to a concerted recruitment effort, was where the Protestants had their greatest success.

Although Company households still retain some economic autonomy (some cultivate their abandoned infields now as extensive outfield plots), the Company in many ways acts as a unit of production and consumption. Company members often work together as a labor group. Put off by the lack of beer, non-Protestants rarely work with Company members. Company members occasionally attend traditional mar muos but drink a nonalcoholic grain beverage prepared for them.

In 1984 I was interested to find that there is also a "company" on the frontier. This is a group of only four compounds, also a Protestant enclave adjacent to their church. It is located in the Kwande ungwa of Mangkogom, which is one of the ungwa that has been depopulated in recent years due to the quality of the soils (chapter 11).

In three separate situations, then, there has been less need for highly concentrated farmstead labor: the reconstructed pioneering phase on the frontier, the depopulated homeland village, and the depopulated frontier ungwa. In each case there are examples of compounds coalescing into larger settlement units. Each case is marked by a greater reliance on communally worked outfield plots than is found on the more crowded frontier today, although in each case, residents of macrocompounds still had to trek to their plots on a regular basis. Thus, although the "pull" to the plot was less, it was still a factor that had to be overridden.

Although I see these relationships among population, intensification, and settlement dispersal as having general applicability, the causes of the aggregation are more particular to the Kofyar case. The first (Doedel's) macrocompound on the frontier had developed a sizable population of Goewanians, and they found themselves acting as a unit and wanting to look after Goewanian interests. The Protestant groups found that as they came to have more in common with each other than with the Catholics and traditionalists, they wanted to become an actual enclave rather than a dispersed subset of the local population.

When conditions of agricultural regime and land tenure do not pull settlements to their land, aggregation may be prompted by relatively weak forces; households may express relations of production by their settlement, may want to cluster with those most like them, or may seek the sociality of aggregation.

Discussion

The dispersal and spacing of Kofyar settlement reflect the simultaneous attraction of the compound to its plot for efficiency of intrafarm movement and to other compounds for efficiency of interfarm movement (Stone 1993b).[8] Kofyar residences are pulled toward their farmland by the need for frequent access to that land, and this is a major factor shaping the pattern of dispersed settlement in the core area. At the same time, reliance on suprahousehold labor in meeting the demands of agricultural production puts a premium on low intersettlement distances.

I have argued that the pull of the residence toward the land caused by intensification results from the proximity-access principle. It is clearly not the case that intensification always produces dispersed farmsteads; farmers may find residence in a nucleated town advantageous if their farmland is fragmented (Bentley 1987), or nucleation may be necessary for defensive reasons (Udo 1965; Rowlands 1972). But the pull remains nonetheless, and there are numerous cases of this phenomenon overriding artificial attempts at nucleation (Gade and Escobar 1982). When there are compelling reasons for nucleated settlement, cultivation of distant plots generally leads to the construction of secondary residences (field houses), or at least the probability of this increases with distance to plot.

Farmstead settlements, then, are closely related to the nature of interaction between sites, as are larger settlements that are involved in trade, purveyance of central functions, and exercise of political control.

9

Agricultural Movement

Except on market days, one is not struck by the movement of people on the Namu frontier. Kofyar farmers spend much of their time in their compounds and adjacent fields, and the actual percentage of their time spent walking or riding more than a short distance is not great. Yet movement is one of the keys to understanding the settlement pattern.

Movement exacts time, energy, and opportunity cost. Attempts to control these costs affect spatial organization in all kinds of economies. The von Thünen theory predicts patterns of land use entirely on the basis of transport cost, Christaller's predictions of central places are based on the distance people will travel to access a given function (the concept of range), and Binford's (1980) model of hunter-gatherer settlement is based on the relative merits of moving goods to consumers and vice versa. Farmers move too, and agricultural movement—defined as movement to or from agricultural plots for purposes directly related to production—has already been shown to affect settlement dispersion. This chapter looks at the spatial structure of agricultural movement, in effect treating farming as a series of movements (Blaikie 1971:4), to contribute to an understanding the relationship between such movement and settlement pattern.

17,000 Trips: Measuring Agricultural Movement

One would expect that agricultural movement has been well studied, but it has not. For instance, despite the breadth of material Chisholm draws on and the rigor with which he builds his case, his book (1962) has a surprising dearth of actual measurements of farmers' movements

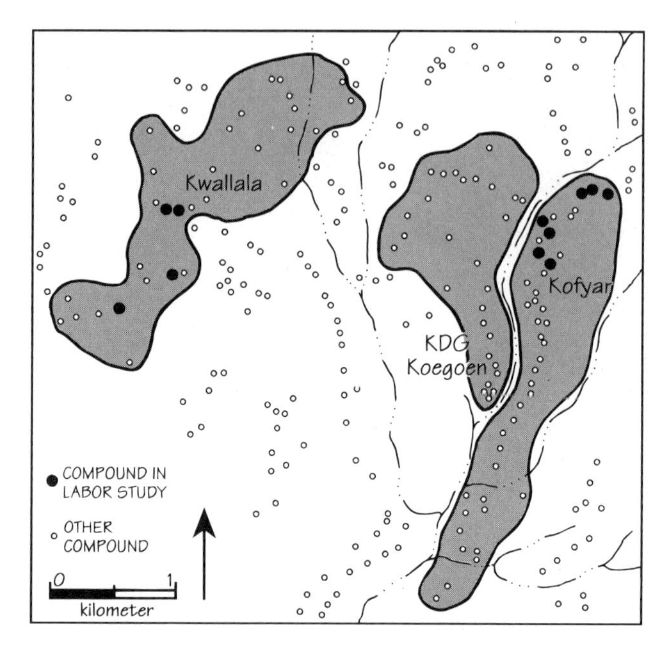

Figure 9.1. Detail of settlement in Goejak and Ungwa Kofyar and the labor study households.

in the conduct of agriculture. Most of his data pertain to agricultural production per hectare and average distance to plots rather than farmers' movements to plots at various distances. Later editions of the book incorporate studies that come closer. The 1979 edition cites Richardson's (1974) estimates of inputs at various distances on Guyanese farms, based on farmers' recollection of how many days they worked particular plots (although the published data are at odds with Chisholm's [1979:44] use of them). Blaikie's (1971) analysis of agriculture in northern India is one of the few studies that records locations of farm operations.

Given the scarcity of real measurements of movement and the scattered nature of the case studies, it is not surprising that Chisholm's conclusions lack firmness. He clearly demonstrates the general decline of labor inputs and agricultural outputs with distance from the residence, which has provided archaeologists with the basis for site catchment analysis (more properly, territorial analysis); but the question of what territory size to use in analysis remains quite open. Archaeologists' territorial analyses have used Chisholm's model to support territorial radii ranging from the 5 km popularized by Vita-Finzi and Higgs to 2 km (summarized in Stone 1991a). Yet resolution of the threshold requires measurements of the spatial dynamics of agricultural systems, which are a logistic nightmare to obtain (Chisholm 1979:34).

By combining information from an agricultural labor study and a settlement survey, I have been able to reconstruct the timing, location, and nature of all agricultural work done by households in the sample group. Figure 9.1 shows the area discussed in this chapter and the locations of the monitored households. Enumerators recorded the type of labor group and the sponsor for all labor parties, and I have identified the provenience of almost all of the off-farm labor bouts. Distances were calculated from the compound of each sample household to every other compound. The location of the residential compound was used as a proxy for the location of the work on other farms; compounds tend to be located centrally within the farm, and workers customarily congregate at the residence before heading to the work site.

A proxy was devised for the own-farm distances, because precise locations of labor bouts on one's own farm are unknown. For the own-farm distances, a series of points was plotted midway between the residential compound and the farm perimeter. Own-farm activities were assigned the average distance from the compound to this midline: midline values ranged from 56 m to 138 m, averaging 84 m.

Eliminating bouts for tasks that were not directly agricultural, such as animal feeding and hut construction, left a sample of 17,066 bouts of own-farm and other-farm agricultural labor. Because I have locations for almost all of the off-farm bouts and a proxy for the on-farm bouts, I can treat bouts as trips to reveal the structure of agricultural movement. I also am able to compare agricultural movement in the two ungwa described in chapter 7: Ungwa Kofyar, with higher population density and intensive cultivation, and Kwallala, with lower population density and less-intensive farming.

Movement Patterns: Territory Size

To get at the question of how far farmers will travel in the normal conduct of agriculture, I first look at how rates of travel decay with distance. To show specifically the frequency with which farmers travel to given distances, I have plotted cumulative percentages of all interfarm agricultural trips against distance from residence in figure 9.2. (This figure excludes the on-farm trips—i.e., trips to a household's own farm—which dwarf interfarm movement; 73.0% of all trips occur within 1 km

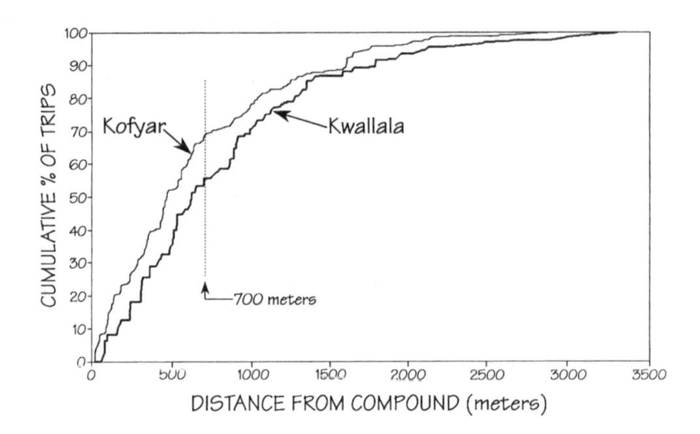

Figure 9.2. Cumulative percentage of trips plotted against distance from compound.

of the residence, and 89.2% occur within the range of the largest home farm. This shows the intensive farmer's need for frequent access to the plots, which has promoted dispersion.) But interfarm movement is a determinant of settlement spacing, and interesting patterns emerge if we focus on interfarm movement. Plotting cumulative percentages of off-farm trips against distance shows that trips increase linearly with distance up to 700 m (fig. 9.2); beyond this, increases in total trips drop off sharply, reaching an asymptote at approximately 2 km. This means that within a range of 700 meters, or approximately 15 minutes of travel time, distance has almost no effect on interfarm agricultural movement. The number of trips increases regularly with distance, the rate of increase being very close to 1% of total trips for every 10 m of distance (i.e., 20% of trips are within 200 m, 50% within 500 m, etc.). When area is made the independent variable, the pattern is similar; when the percentage of trips is plotted against a circular area centered on the residence, the slope of the line is of course lower, but the threshold at which trip percentages begin to taper off is 130 ha, the area within 643 m of the compound.

These findings show a clear threshold below which distance does not affect farmers' willingness to travel from their farmsteads. The finding of such a threshold, with the sharp drop-off in inputs beyond it, agrees with the general structure of agricultural movements proposed by Chisholm, but the threshold radius is actually measured at 700 m rather than estimated at 1 km. It also must be kept in mind that the majority

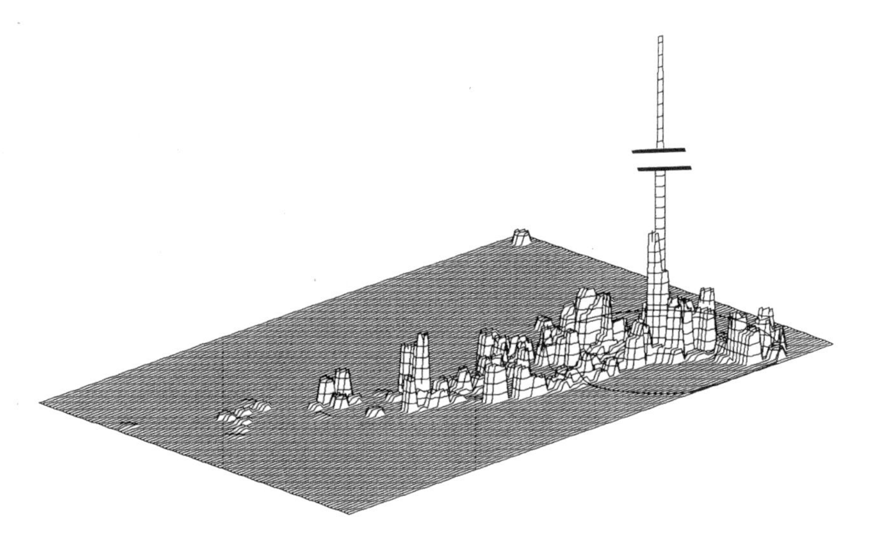

Figure 9.3. Aggregate agricultural movement in Ungwa Kofyar. The per capita trips made from the seven monitored households in Ungwa Kofyar to each point on the landscape relative to the household's residence were combined into one data set. This data set was then used to create a topographic surface, with elevations calculated by averaging each point with its nearest neighbor within a search radius of 50 m. The scale for the z-axis is shown by the horizontal lines. The elevation of the peak at the origin is 391.

of agricultural work occurs on one's own farm, and the distance to this work is kept extremely low by settlement dispersal.

We can also isolate the effect of agricultural intensity on movement patterns. The numbers of off-farm agricultural trips are identical (59), but Kofyar farmers put in 21% more trips to their own plots than did Kwallala farmers (391 vs. 324). The longer workdays (18.2% more labor inputs) and lower marginal returns in Ungwa Kofyar mean that time there is at a premium. Because farmers in Ungwa Kofyar are under greater scheduling pressure they might be expected to minimize travel costs more than in Kwallala. Figure 9.2 also compares the decay of interfarm travel in the two ungwa. The difference in cumulative percentages is greater than is revealed at first glance; for instance, the radius mark at 475 m contains more than half of the trips in Ungwa Kofyar but less than a third of the trips in Kwallala. The ogives show that despite the difference in distance decay, both movement patterns show the slope change at about 700 m.

Territory Shape

Analysis of archaeological site territories also requires assumptions about catchment shape. Researchers have assumed territories to be square, circular, or isochronic (Stone 1991a). On the hypothetical isotropic plain, the result of distance or time expenditure should be the concentric pattern described by von Thünen. In reality, social factors can override the effects of sheer distance, although this process is hard to isolate.

I have depicted the spatial organization of agricultural movement using three-dimensional "mountain-range" graphs in figures 9.3–9.6. For each household, the total number of trips to each x,y coordinate on the landscape is calculated and divided by the number of adults in the household to yield per capita trips. To see the aggregate pattern of agricultural movement, I combine all the individual mountain-range charts for an ungwa, with the origins aligned over the same point. Movement can then be displayed in a mountain-range chart in which the height of the "peak" shows the summed per capita trips to each location. The circle indicates a radius of 700 m (the recurrent threshold in the Kofyar analysis).

Figures 9.3 and 9.4 show aggregate agricultural movement for Ungwa Kofyar and Kwallala. The pattern of distance decay is immediately apparent in the density of oft-visited locations near the home. But in both ungwa, the shape of the movement pattern has irregularities. If the movement data were based on a large random sample of farm households, the decay would presumably be the same in all directions. But it is not, and the shape of the movement reveals much about how the social organization of production affects settlement pattern.

To understand the shape of movement territories we have to consider the interplay between two essential principles of Kofyar agriculture that both agree and compete. The first is the movement cost discussed above; beyond the threshold of 700 m, the friction of distance constrains agricultural movement. This means that there is not only a premium on exchanging labor with compounds within 700 m, but a cost to not exchanging labor with compounds within that range. (The logic is that a dependable group-labor system requires numerous participating households, and the fewer of those households that are not within the 700 m range, the more must be beyond it.) With the Kofyar, then, the principle of controlling movement costs promotes travel to as many

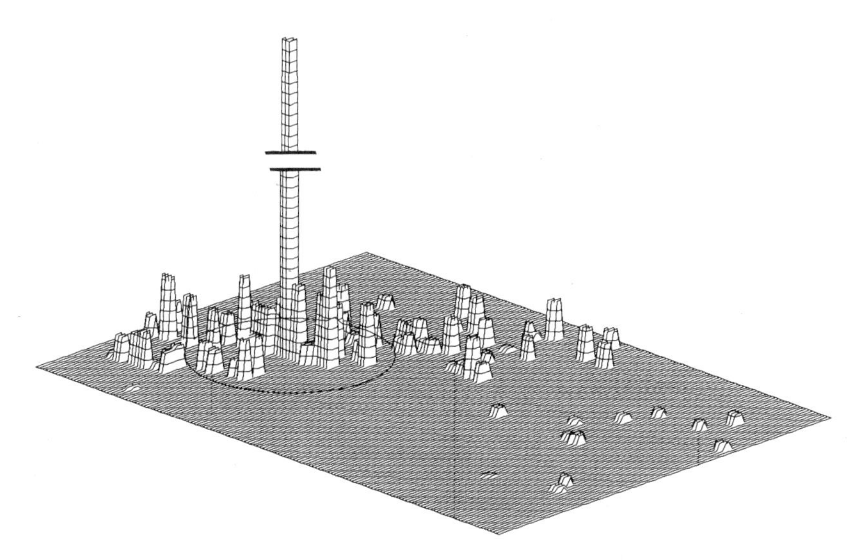

Figure 9.4. Aggregate agricultural movement in Kwallala. This is based on the four monitored households in Kwallala, using the method described for figure 9.3. The elevation of the peak at the origin is 324.

farms within 700 m as possible, with further travel decreasing with distance.

The second principle is the division of the landscape into ungwa, which are forums for labor mobilization. Ungwa are the key to the dependability and coordination of the labor supply. Yet the same segmentation that makes group labor dependable also conflicts with the principle of movement cost control; in other words, the social aspects of agriculture contradict the spatial aspects of it.[1] To see how the conflict is resolved, we can look first at how ungwa boundaries do affect movement, and then at how they are overridden.

Figure 9.3 shows the effect of ungwa boundaries on agricultural movement in Ungwa Kofyar, where the deformation of the concentric model is especially evident. The confinement of the overwhelming majority of agricultural movement to within ungwa boundaries is especially notable because of Ungwa Kofyar's elongated shape and the location of the sample farmers near the northern end, which leaves many of the areas that lie within 700 m outside of the ungwa. At the same time, workers are pulled well beyond the 700 m threshold to the south-

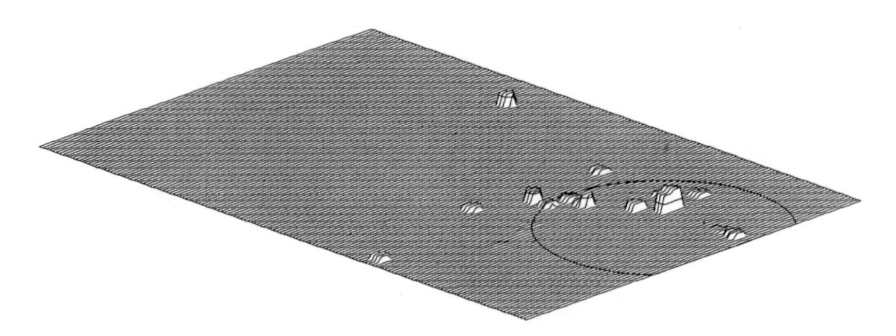

Figure 9.5. Agricultural trips across ungwa boundaries. The trips shown are only from Ungwa Kofyar, since none of the recorded trips by Kwallala farmers was outside of their ungwa. The graph was produced by the method described for figure 9.3.

ern end of the ungwa, something they would rarely do except to contribute to the communal labor system.

Goejak is especially interesting in this regard because of its division into noncontiguous sections. Figure 9.4 clearly shows the travel of Kwallala residents to the KDG Koegoen section, despite the intervening portion of Rafin Gwaska and the sheer distance (see fig. 6.2). Meanwhile, the farmers monitored in Kwallala did not attend a single labor party in Rafin Gwaska; in fact, they did not attend a single labor party outside of Goejak.

Ungwa boundaries constrain aggregate agricultural movement to a great extent. The pivotal role of the ungwa in the social and spatial organization of agriculture is summarized in the remarkable statistic that more than 98% of the 17,066 recorded agricultural trips were made within ungwa boundaries (97.1% of the 1537 interfarm trips and 99.6% of the 10,868 total trips by Kofyar farmers; 96.3% of the 1079 interfarm trips and 99.4% of the 6198 total trips by Kwallala farmers). Ungwa shapes can therefore prevent evenly concentric movement by both pulling workers to relatively distant farms and deterring movement to nearby farms that are across ungwa boundaries.

But how strongly is movement controlled by the social organization of production as compared to simple movement costs? The Goejak farmers in our study did not make a single trip outside the ungwa, although there was movement between the separated subareas described above. But there was interungwa movement by Ungwa Kofyar farmers, the shape of which is depicted in figure 9.5. A little movement occurs

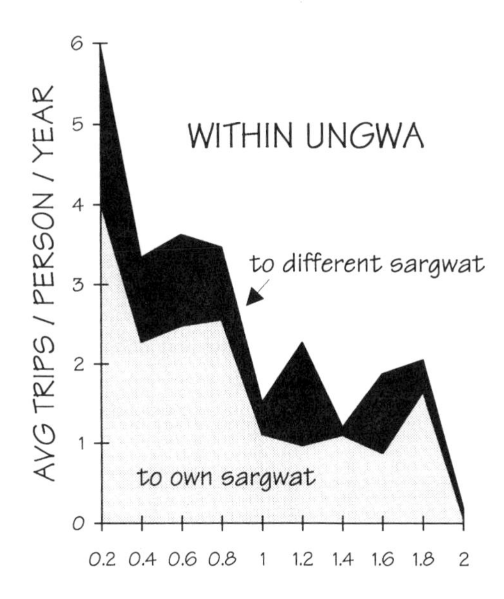

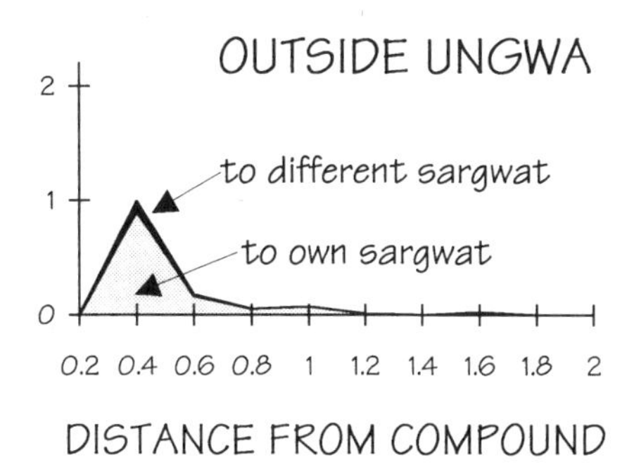

Figure 9.6. Drop-off of agricultural movement.

across the northern boundary into Wunze, but only up to the 700 m threshold. Most of the trips were to KDG Koegoen, where much of the 700 m ranges lay. Of the interungwa trips, 63% were within 700 m, and 83% were within 1 km.

Figure 9.6 directly compares movement within and between ungwa for Ungwa Kofyar.[2] The lines show, for various distances, what percentage of all farms in and out of the ungwa were visited at least once for farmwork by our sample. It shows, first, the strong pull to the farms of ungwa mates. Well over 90% of the farms within 1 km were visited;

beyond that the percentages drop off, but even in the 1800–2000 m range—approximately a 30-minute walk—almost two-thirds of the ungwa's farms were being visited. The overall numbers of off-ungwa visits are minuscule in comparison, although the chances of a visit do climb abruptly within the 700 m radius. There were no off-ungwa farms within 200 m of our sample households in Ungwa Kofyar, but one-third of the off-ungwa farms within the 400 m range were visited.

It should be clear that the spatial organization of Kofyar agriculture cannot be understood only as the patterns emerging from individual farmers confronting von Thünen's friction of distance. It is rather a social process, with a vital element of interdependence in mobilizing the labor that drives production, which promotes sociosettlement units to provide dependability of labor exchange. Figure 9.6 demonstrates how agricultural movement is shaped by these sociosettlement units (ungwa) at the same time that individual farmers prudently regulate their travel-time costs. When these two priorities conflict, the farmers compromise: they are willing to travel even to distant farms in their ungwa, but they allow their participation to wane beyond the 1 km radius; they also mainly confine their labor exchange to within ungwa boundaries, although those boundaries are increasingly ignored as distances fall within the 700 m threshold.

In treating agricultural movement as being shaped by work-exchange territories, I am directly evoking archaeological research on site territories (catchments), and in particular work such as Flannery's (1976a:178), which supposed that social reasons, "perhaps measured in travel time," were overriding the purely spatial considerations in farming. I have shown how social factors that are a fundamental component of production—the ungwa structure—can both override and be overridden by the effects of distance. Let us next consider the effects of another social factor—the patterns of affiliation brought from the homeland—on agricultural movement.

Spatial and Social Propinquity

One might suspect that if Kofyar farmers are stingy with trips across the ungwa, there may be other social factors affecting trip decisions, and there are. In fact, there are strong relationships between agricultural movement patterns and the "microethnic" relations described in chapter 4. What complicates matters is that this social propinquity affects not only how far one will travel to other farms in a given settlement pattern, but how farms are located within the settlement pattern in the first place, as well as how ungwa boundaries are drawn. I have previously described how sargwat affiliation shaped early migration patterns and how village affiliation affected ungwa formation (chapters 5–6). In this chapter I am simply treating the ethnic settlement pattern as a given, postponing the analysis of the effect of ethnicity on settlement location for chapter 10.

In chapter 4 I argued that "ethnicity" among the Kofyar was hierarchical and complex, including some important social taxa that lack characteristics often emphasized in contemporary research on ethnicity, such as self-concept. Yet culturally and behaviorally distinct groups may indeed lack an inclusive name and self-concept, and the Kofyar warfare alliances (sargwat) turn out to be units of social propinquity with wide effects on behavior.

Figure 9.6 graphs the average number of trips each sampled worker was making per year to farms in and out of his/her home ungwa, to farms of members of his/her own sargwat versus other sargwat. It shows that frequent travel is not only largely restricted to the ungwa and to the 800 m radius (the breakdown does not show the 700 m radius), but to sargwat mates, whose farms are visited nearly twice as frequently as others. Especially striking is the picture of interungwa travel, which is virtually always to sargwat mates; even within 400 m, the rate of travel to non–sargwat mates across ungwa boundaries is less than one-tenth of a trip per person, whereas the rate of travel to sargwat mates within this range (but across ungwa boundaries) is 10 times greater. In this way, the "deep" social affiliations from the homeland override the frontier labor pools, but only within ranges as close as 600–700 m.[3]

10

Ethnicity and Settlement

The issue of how settlement patterns are affected by ethnic relations has now been broached several times, and it may already be apparent that this relationship is an elaborate one. There is first the problem of defining the ethnic taxonomy brought to the frontier, and beyond that the question of how and why any of the levels of affiliation described in chapter 4 should influence settlement location. There is then the complicating fact that causality runs both ways between social and spatial proximity: the socially close tend to settle near each other on the frontier (as we will see), and those who settle near others often tend to become socially close.

There is also a lack of theoretical writing on the subject; there are good case studies of ethnic settlement but few attempts to explain the cases at a general level. As I outline in chapter 2, a tendency for social propinquity to be expressed in spatial proximity has almost been taken for granted, the logical result of kinship (Jordan 1976; Bohland 1970) or regional consciousness (Lehr 1985). In trying to make some sense of Kofyar microethnicity and settlement, it is with this general theoretical problem of the relationship between social and spatial propinquity, viewed in the production-oriented perspective, that this chapter begins. I am especially interested in the effects of ethnicity on settlement on the frontier, both because this allows us to partly de-historicize the issue and because our case study is set on a frontier.

Social and Spatial Propinquity on the Frontier

To historian Frederick Jackson Turner (1920:23), the American frontier was a leveler of ethnic differences, a crucible in which "the immigrants were Americanized, liberated, and fused into a mixed race" (also see Mikesell 1960:62; Ostegren 1979:189; Rice 1977). As several scholars have pointed out (Hudson 1977:24; Lehr and Katz 1995:413), this has never squared very well with the perspective in cultural geography and anthropology, where the emphasis is less on the transforming power of the frontier than on how frontier societies reflect traits of the core (e.g., Kopytoff 1987). Although some have complained that the "Turner hypothesis" was too vague to be a hypothesis (Savage and Thompson 1979), it was at least hypothetical in that many of its tenets had yet to be investigated empirically. Since Turner's day, a lot of research on New World frontiers has illuminated the role of social propinquity in settlement patterns. A recurrent theme in this research is that, far from being the admixtures of culturally diverse individualists Turner envisioned, settlement patterns are strongly shaped by patterns of social affiliation at levels ranging from nationality to "ethnicity," province, parish, village, family (Eidt 1971; McQuillan 1978; Brunger 1982), and even the boat on which immigrants arrived (Brunger 1975).

Many of the most revealing studies have focused on the historic geography of the North American West; the pull of social proximity on settlement appears consistently, whether with Russians or Swedes in Kansas or Minnesota (McQuillan 1978; Rice 1977), Mennonites in Saskatchewan (Friesen 1977), or Ukrainians in Alberta (Lehr 1985). An interesting variant on this pattern is the Mormon settlers who arrived in groups already having a map of their yet unbuilt gridiron town (Katz and Lehr 1991). Contrary to Turner's frontier where the pioneer escaped old communal ties, studies on the North American West and Midwest have consistently shown a pull that, in the terms developed in chapter 3, we would call ethnic gravity. The key to drawing any general lessons from such patterning is determining why this gravity occurs. Let us deal with this general issue before turning to analysis of the Kofyar case.

The Economics of Attraction and Affection

Chapter 2 discussed how kin-based clustering of small settlements is often seen as an artifact of the "economy of affection" (Hyden 1980:18–19 and passim). The economy of affection is often contrasted to the economy of production, as with the farmers on a Philippine frontier who spurned the most productive land because of a "noneconomic criterion, such as proximity to kinsmen or friends" (Eder 1982). Alternatives to models where individuals "maximize pecuniary functions of cost, profit and utility" have been offered that consider "the degree of satisfaction derived not only from pecuniary advantages but also from non-pecuniary utilities such as interaction among group members and activity carried on with the groups" (Gjerde 1979:405).

No one would deny the importance of these nonpecuniary interactions, but we systematically tend to underestimate the economic importance of neighbors, biased perhaps by our life in settlements where arrangements of most residences are independent of (or only loosely and indirectly related to) one's productive activities. Yet the psychological forces of attraction often overlap the economic functions of kinship and social affiliation, as reflected in the Kofyar homeland: "All things being equal, a young man would probably choose to live near his brothers and other lineage mates. They can be expected to make land available to him most readily and protect his interests most assiduously. *He can trust them to aid him in the fields,* to support him in most disputes and fights, and to consult with him on political matters. These advantages are enhanced by nearness of residence" (Netting 1968:147–148, emphasis added). I am concerned not just with the advantages of "communication, convenience, and neighborly help" (McQuillan 1978:141) allowed by ethnic clustering, but with the mobilization of aid in the fields as a factor promoting ethnic gravity. Our images of the rugged individualist on the frontier and the self-reliant farmer of today notwithstanding, smallholders virtually everywhere rely on the labor of neighbors, even though the household may be the main unit of production (Netting 1993:194 and passim).

But however dependent the farmers may be on labor pooling, they are wary of it as well. For the leverage provided by access to labor groups, the farmer pays a cost in control over his or her own labor schedule and risks a net loss in labor mobilized for his or her own farm. Farmers take seriously control over their own schedules, and they assess their net labor costs carefully, especially as agriculture is intensified. The more farmers tend crops they will neither consume nor sell, the greater the risk of serious disagreements that threaten the dependability of the labor system. It is imperative that cooperative labor be dependable, and it is unlikely to be dependable unless the participants share pivotal concepts of reciprocity and work; mutual aid "depends on the ability to manipulate relevant symbols and participation in noneconomic social and cultural activities" (Williams 1977:77).

This helps explain why it is so common for work groups to comprise members of the same ethnic units (Erasmus 1956:446), such as the Michigan men who banded together in cooperative activity around the planting, harvesting, threshing, and storing of farm products: "In places where members of different ethnic groups are living together it was found that the Polish farmers who might have silo-filling or threshing groups included only those of Polish ancestry, while farmers of old American background similarly grouped among themselves" (Kimball 1950:38). It also explains Erasmus's important observation, "Both types of work party [exchange and festive] are made up of relatives and/or neighbors; kinship, friendship, *and proximity* tend to overlap in the rural societies which practice reciprocal labor" (Erasmus 1956:446, emphasis added).

In chapter 8 I showed both in theory and with the Kofyar data how agricultural movement directly affects settlement spacing and farm shape. Consideration of the social aspect of cooperative labor shows an important element in ethnic gravity; there is an advantage not only in adjusting settlement to allow efficient movement to nearby farms, but in having those nearby farms inhabited by agricultural collaborators. Because agricultural collaborators tend to be ethnically close, this is an agroecological basis for ethnic gravity in agrarian settlement. There is more at stake than the mere desire to be near kin, or McQuillan's "communication, convenience, and neighborly help"; it is directly related to the social organization of agricultural production.

The economic basis of ethnic gravity is reflected in the relative success of agrarian communities sorted out so as to facilitate agricultural collaboration. Although the patterns are somewhat equivocal, various scholars have noted a greater tendency for frontier communities to prosper when their members are socially close, whether the settlers were bound by kinship, home village affiliation, language, or some aspect of ethnicity (Brunger 1982). Rice (1977:171–172) found that persistence of settlement in 19th-century Minnesota was related to community cohesion, with those coming from the most diverse provinces in Sweden having lower persistence rates. Eidt's study of agricultural colonization in Argentina shows that pioneers sorted themselves out in general by nationality and that communities with settlers from the same homeland communities were the most successful (1971:105–115). He found that the progress of colonies settled by pioneers from the same home communities was "an understandable phenomenon because old neighbors trusted one another and worked together from the start." Where residence patterns promoted mutual trust, agricultural cooperatives formed that were invaluable in tasks with simultaneous labor demands. These conditions also facilitate the management of common property.

Especially on the frontier, there can be distinct advantages to living not only near those who are socially close, but in distinct territories "within which cooperation and reassurance prevailed" (Brunger 1982). Suttles (1972) describes principles of urban territorial groups that apply to the rural: recognized territories designate the range of trustworthy associations and build accountability. Formal recognition, naming, and demarcation also help establish control over land (Stone 1994).

There are, then, ecological bases for social propinquity's playing a significant role in the determination of frontier settlement location, a role based not merely on affection for kin, friends, and co-villagers, but on facilitation of the suprahousehold organization of production. The question is how different levels of social affiliation or propinquity affect the social organization of production and thus settlement.

One of the issues that interested me most on the Kofyar frontier was how ethnic gravity was manifested—which levels of ethnicity were germane to land use, and how the relationship changed with agricultural intensification. We will begin by exploring the microethnic composition of ungwa.

Ethnic Makeup of Ungwa

I have described broad differences in settlement in the early-frontier spread of bush farms. Doemak and Kwalla farmers spread mostly to the south, whereas many Merniangs struck out to the east, across the basalt flows and toward the area between Doka and Shendam. There were Merniang farms on the southern piedmont as well. I do not know the extent to which these early bush farms were segregated by tribe or sargwat, but the sedimentary soils of the Namu Plains were reached by both Merniangs and Doemaks soon after 1950. It is clear that early settlement of the Namu Plains was segregated by tribe; my focus is on the Merniang area (chapter 5).

In 1962, Netting visited the fledgling community, collecting 99 short censuses and recording home villages for the household heads in five ungwa: Ungwa Long, Wunze, Bubuak, Ungwa Goewan, and Hanyar Kwari. These were predominantly Gankogom/Merniang ungwa; other ungwa to the west were mostly Doemak. Figure 10.1 shows a recon-

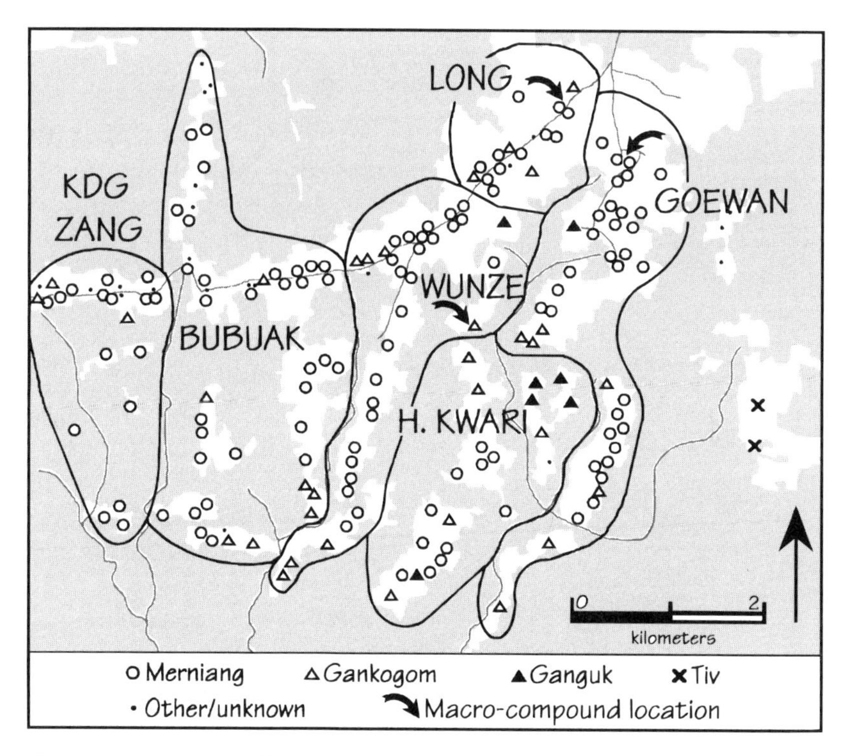

Figure 10.1. Microethnic composition of early frontier core-area ungwa, 1963. The map is based on analysis of aerial photographs, interviews, and Netting's field notes. Although Netting recorded 194 households within these ungwa, only 174 compounds appear on the aerial photographs because some compounds held more than one household.

struction of the microethnic settlement pattern on this early stage of the frontier.

Let us first look at the ethnic makeup of the groups that considered themselves ungwa. Pioneers from hill villages show a moderate tendency toward clustering in particular ungwa; all 13 Kopfuboem households are in one ungwa, almost all 17 Pangkurum households are in one ungwa, and the 18 Bogalong households occur in only two ungwa. Households from other villages, like Kofyar, Longsel, and Gogot, are more spread out. In general the Merniang villages are much less concentrated (in the later section on settlement encystment I discuss some reasons why this would be). Figure 10.1 shows almost no patterning of ungwa composition by sargwat. The Gankogom-Merniang percentages in the ungwa are fairly constant, with a few Ganguk households in addition. Settlement began in Hanyar Kwari somewhat later than in the other ungwa, and by 1961 it had attracted a group that was somewhat more ethnically diverse.

Why, then, if the Doemak sargwat went its own way in settling the Koprume area, was there a surprising amount of intermingling of the Gankogom, Merniang, and Ganguk sargwat elsewhere in the core area? I believe it was in large part an artifact of events surrounding the pioneering phase. I have pointed out that a viable colonization of the Namu Plains required sufficient numbers to provide for labor pooling, scare off animals, and retain the land against outsiders. Doedel, chief of Kwa, had joined forces with Baam (chief of Lardang) as part of his campaign to raise a viable pioneering community. Both Doedel and Baam had been part of the initial compound cluster near the north end of what is now called Ungwa Long. The union of the Lardang and Kwa chiefs that was contrived to spur migration crosscut the sargwat boundary, which activated migratory streams from both Gankogom and Merniang into the same part of the frontier.

Yet even among the early pioneers, there was some separation of farm locations by sargwat. The fission, soon after, of the group of village mates who formed Goewan gari on the other side of the river was a reassertion of ethnic segregation. The 15 Pangkurum households in Wunze were set off from the beginning, the result of a communal meeting and decision to move to the frontier as a group. Hanyar Kwari's four Ganguk households were also on contiguous farms.

It was not long before the Gankogom and Merniang migratory streams began to diverge more obviously. Even at the time of Netting's census, Gankogom migrants were filtering into their own areas. A common pattern was to live at first with an established Gankogom farmer and then move into what is now Ungwa Kofyar or KDG Koegoen.

By the 1980s, sargwat mixing in most ungwa had been definitely reduced. Figure 10.2 graphically depicts the ethnic breakdown for household heads in the 25 ungwa for which I have reliable information. Broad patterns, some of which have already been described, are immediately apparent: Doemak farmers are predominant in Koprume, on the piedmont north of Namu, and in areas east of Namu, whereas Gankogom and Merniang farmers dominate this study's core area.

Table 10.1 contains the data on which figure 10.2 is based, showing the extent to which ungwa are segregated along microethnic lines. If we say that an ungwa is "dominated" by a sargwat if 70% or more of the household heads are from that sargwat, then 18 of the 32 ungwa are dominated by a single sargwat. The Merniang and Gankogom sargwat together dominate another 9 ungwa. The tendency for these groups to settle together reflects their social proximity as reflected in marriage patterns (and, less importantly, in their together forming the Merniang tribe). The association also reflects the history of settlement, which initiated migratory "streams" into the same area.

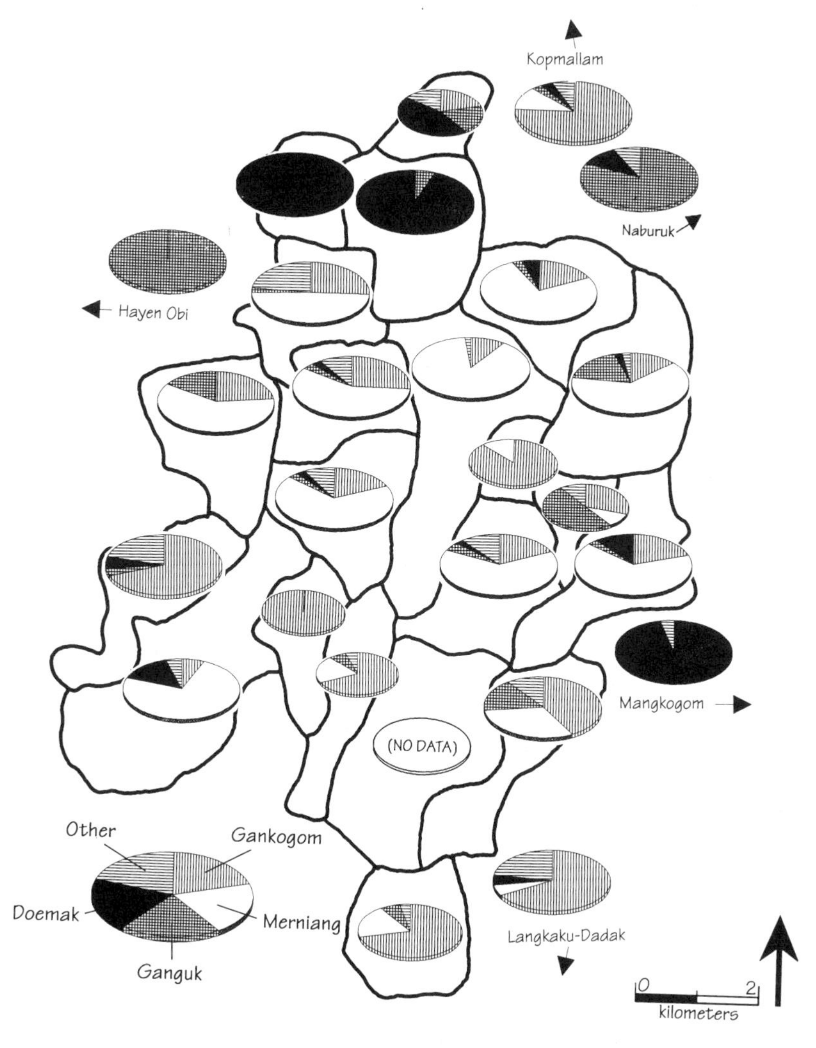

Figure 10.2. Microethnic composition of frontier core-area ungwa, 1984. Data are unavailable for Burugu.

Table 10.1 Makeup of 1984 Ungwa by Sargwat of Household Head (row percentages)

	n	Jipal	Gankogom	Merniang	Ganguk	Doemak	Mwahavul	Other
Doemak								
Kopmoekam	20	—	—	—	—	100	—	—
Koprume-Dakup	39	—	—	—	3	97	—	—
Mangkogom	23	—	—	—	—	96	—	4
Doemak (with non-Kofyar)								
Kopdayim-Dasoem	27	—	—	—	—	63	11	26
Dadin Kowa	26	—	—	3	4	74	2	17
Kopdayim-Dashagal	23	—	—	—	—	70	30	—
Merniang								
Dawam	9	—	11	89	—	—	—	—
Wunze	64	2	9	88	—	—	—	2
Merniang-Gankogom								
Kangiwa	22	—	73	18	5	5	—	—
Duwe North	36	—	22	64	3	11	—	—
Kopmallam	32	—	75	13	3	3	3	3
Hanyar Kwari	49	—	14	61	4	2	4	14
Bubuak	34	3	26	59	3	3	3	3
Kofyar	43	—	67	21	7	—	—	5
Long	35	—	17	74	3	6	—	—
Dunglong	34	—	18	71	3	—	—	9
KDG Zang	44	—	23	61	16	—	—	—

Merniang-Gankogom with Minority Contingent								
Duwe South	38	3	42	29	13	—	—	13
Goewan	46	—	13	63	17	2	—	4
Kopdogo	39	—	26	49	3	—	23	—
Rafin Gwaska	66	—	2	74	2	—	5	18
Gankogom								
KDG Koegoen	25	—	100	—	—	—	—	—
Pangkurum	8	—	88	13	—	—	—	—
Dangka	30	—	83	7	7	—	—	3
Ganguk								
Mandeshik	9	—	—	—	100	—	—	—
Naburuk	27	—	—	—	81	11	—	7
Hayen Obi	9	—	—	—	100	—	—	—
Mixed								
Koprume-Dayim	32	—	13	—	9	53	3	22
Hayen Akuni	15	—	27	7	20	27	—	20
Gogot	20	10	30	10	50	—	—	—
Kwallala	29	17	66	—	3	7	3	3
Langkaku-Dadak	18	11	67	6	—	6	11	—

Note: Households are assigned microethnic (sargwat) affiliation according to the home village of the head because heads of households normally make locational decisions. The head's sargwat is a strong indicator of wives' sargwat; in the sample of 1496 wives in frontier-farming households, 72% are married to a man from the same sargwat.

Mechanisms of Microethnic Segregation

How does the degree of microethnic segregation we see in the core area come about, and how is it maintained? Some of it comes about by social propinquity, in the channeling of the flow of information about land availability. Information is a vital resource that must be gathered directly through observation or indirectly through social processes of information distribution (Moore 1983:178). Crucial information on land availability on the frontier tended to follow lines of social propinquity in the homeland, so that a farmer was most likely to know about available land in areas being settled by village mates. This is why "once a migratory stream begins pumping in a given direction it continues to flow in that direction" (Lefferts 1977:39).

In some situations the early migrants do more than simply convey information about land availability, including actively encouraging others to join them. This was certainly the case in the pioneering phase, when the first arrivals entreated others to add to their labor pool and help claim the land from the animals. The situation paralleled much longer range migrations to underpopulated areas, such as when early European immigrants on the Great Plains of the United States sent home "America letters" entreating others to join them and sometimes even including prepaid tickets (Gjerde 1979:406). Because these pleas were directed to kin and friends, they contributed to the spatial replication of social distances on the frontier.

A case in point was a Kofyar mengwa near Langkaku. Facing declining yields on his farm near Kurgwi, he was looking around for a new farm south of the piedmont in the 1960s. He had been offered a plot near Dangka, but he found it too small for his family; he then heard that land might be available near Langkaku. There were no other Kofyar there at the time, but land was plentiful and he got several others to start farming there. Over the next several years he encouraged others to move there to make the ungwa viable. He in effect admitted to me that this required some effort, because the Langkaku farmers began to discover problems with the soils. In describing the soil to others from his home village, he would tell them, "Well, *I* find it good" (*a kat dong*) rather than the usual perfunctory "It's very good" (*dong sosei*). By around 1980 his ungwa had grown to 30 households, mostly from his home village of Bogalong, but this number had been nearly cut in half by abandonments by 1985.

There are also ways in which social propinquity in the ungwa is actively controlled by its residents. There is a pattern of mengwa offering available plots to co-villagers or co–sargwat members. The mengwa have titular control over the plots, but in fact few have the authority to override community consensus, and a land seeker is much more likely to be allowed into an ungwa if he has links in it. For instance, there was a small plot in Ungwa Kofyar where a non-Gankogom was being allowed to plant crops. However, although the owner had a surfeit of land, the owner would neither sell the land nor allow the outsider to build a field house there, for fear it would constitute a small step toward possessing the land. By contrast, although we had been told several times that there was room for no more farms in Ungwa Kofyar, we saw just before we left the field a new compound being erected for a young man from Kofyar village. More severe is the tactic of driving out farmers who have managed to acquire land in an ungwa. I know of just a few cases of this, but the expelled farmer tends to be an ethnic outsider.

Microethnic segregation is also achieved by adjusting ungwa borders. In the early 1960s, when pioneers from Kofyar village were settling at the southern end of Wunze, they informally named the locality after their home village.[1] Later, they began to refer to it as an ungwa in its own right, but their taxes were submitted through the mengwa of Wunze before taxes were abolished in the early 1980s. When the head tax was reinstated in 1984, Ungwa Kofyar paid on its own, legitimating its separate status.

Goejak is a rarity: a discontinuous ungwa, comprising sections called Koedoegoer Koegoen[2] and Kwallala. The settlers in both sections are predominantly from Lardang (table 10.2) and so considered themselves a single ungwa despite the intrusion of the northern end of the Merniang-dominated ungwa of Rafin Gwaska.

Toenglu in Frontier Ungwa

Ungwa in the core area are divided into neighborhoods, typically comprising around a dozen farmsteads each. These neighborhoods parallel the neighborhoods in homeland hill villages, and they are called by the same name: toenglu. The toenglu acts as a forum for mobilizing reciprocal wuk labor just as the ungwa is a forum for mobilizing mar muos.

Figure 10.3a shows the toenglu divisions in Ungwa Kofyar; the boundaries can obviously delimit areas of social propinquity (note the toenglu called Kongo). Figure 10.3a also shows the locations of wuk memberships. These are household wuk, meaning that they work on the household fields of the members, and actual workers at any meeting can be any adult representative of the household. At least as of 1985, all household wuk groups were organized within toenglu boundaries, and in two cases the wuk membership included the entire toenglu.

There are also nonhousehold wuk groups; in these, the members are individuals, and the work is done on the individuals' plots. These are usually age- and sex-specific, and they often cross toenglu boundaries to get their membership to desired levels. Figure 10.3b shows the locations of members of a young men's wuk that crosscuts two toenglu and of the members of an all-women's wuk group. The women's group, which called itself Gongdogom, was more formal than some wuk, having its own president, vice-president, and treasurer (M. P. Stone 1988b:67). This group's membership crosscut three toenglu.

Ethnicity and Site Locations: Encystments and Nearest Neighbors

We have looked at the dynamics of ethnicity at the ungwa and toenglu levels, but some of the most distinctive effects of ethnicity on settlement pattern appear in the arrangements of individual residences.

Figure 10.4 shows the microethnicity (by sargwat) of 862 household heads in the core area. It shows fairly distinct areas dominated by particular sargwat, but it also shows that these domains are not pure in composition. Where domains are intruded upon there is a distinctive pattern of encystment by the minority group: the greater the social distance between the dominant group and the intruders, the more strongly the intruders are attracted to each other. Strong encystment is shown at the inclusive tribal level in Kopdogo, where 6 Mwahavul farmsteads are grouped together, and in Koprume where there are 8 Ampang Mwahavul farms in a tight cluster. Encystment at the sargwat level occurs most clearly in Ungwa Goewan and Hanyar Kwari, where a cluster of 6 and 10 Ganguk farms are set off from the dominant Merniang and Gankogom farmsteads. There is a encystment of 6 Pangkurum and 2 other Gankogom farmsteads in the Merniang-dominated ungwa of Wunze. In northeastern Koprume-Dayim is an encystment of Gankogom farms in a Doemak domain, and in southern KDG Zang an encystment of Doemak farms abutting domains dominated by Gankogom and Merniang. Figure 10.5 isolates the two most clearly encysted groups in the core area: the Ganguk sargwat and the Mwahavul tribe.

It would be helpful to quantify the attraction to settlements of the

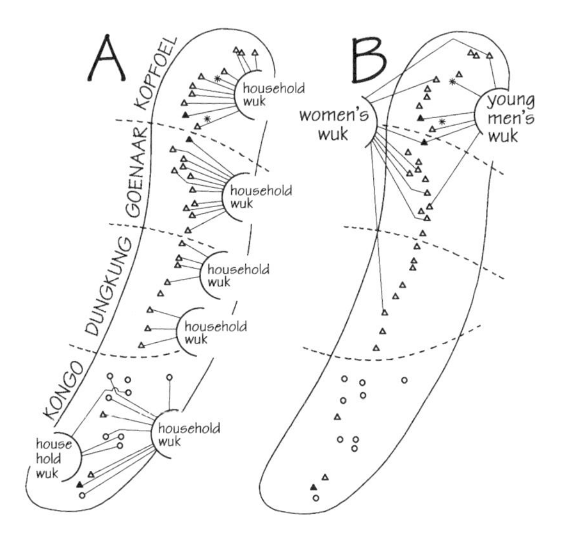

Figure 10.3. Toenglu and work organization in Ungwa Kofyar, 1984.

various levels of social distance. The oft-used nearest-neighbor statistic does not recognize different classes of points within the point pattern, and quadrat-based approaches to spatial segregation (e.g., White 1983) are too coarse for these purposes. My question is whether farmsteads are attracted to each other by shared village or sargwat affiliation, which can be measured by looking at the village and alliance affiliation of each farmstead's two closest neighbors.[3] Table 10.2 shows this for villages represented by at least 10 farmsteads in the study area. It lists the percentage of farmsteads that have at least one co-villager as a nearest or second-nearest neighbor. It also shows the binomial probability of this happening if locations were random with respect to village affiliation. It then adjusts the actual percentages by dividing them by the binomial probabilities, yielding an attraction coefficient.

Table 10.2 shows farmsteads to be generally but not overwhelmingly attracted to co-villagers. Given the many factors that may affect settlement location, I will not belabor minor differences in attraction coefficient, but the major differences are notable. The strong attraction among Ampang and Dung settlers demonstrates the encystment principle discussed earlier in this chapter. Ampang farmers are Mwahavul, differing from their neighbors at the level of the inclusive tribe; most Dung settlers are in an area dominated by the socially distant Merniang sargwat. The villages with the lowest attraction coefficients tend to be

Table 10.2 Village and Sargwat Affiliations of Nearest Neighbors

	n[a]	Proportion of Sites with Co-Villager[b]	Probability of Another Site Being Co-Villager[c]	Binomial Probability of a Co-Villager[d]	Attraction Coefficient[e]
Home Village					
Ampang (Mwahavul)	11	0.73	0.01	0.03	27.2
Dung	10	0.60	0.01	0.02	24.9
Dunglong, Kwa	20	0.60	0.03	0.05	11.9
Miket	23	0.52	0.03	0.06	8.9
Kofyar	31	0.65	0.04	0.08	8.1
Kwanoeng	30	0.60	0.04	0.08	7.8
Pangkurum	13	0.23	0.02	0.03	7.2
Meer, Kwa	15	0.27	0.02	0.04	7.1
Kongde, Lardang	11	0.18	0.01	0.03	6.8
Chinwaar, Kwa	22	0.36	0.03	0.06	6.5
Gatoeng, Lardang	36	0.53	0.05	0.09	5.7
Kwang	12	0.17	0.01	0.03	5.7
Goewan, Kwa	45	0.58	0.06	0.12	5.0
Fogol, Kwa	60	0.52	0.08	0.15	3.4
Lardang*	12	0.17	0.01	0.03	5.7
Doemak*	68	0.79	0.09	0.17	4.6

Kwa*	45	0.33	0.06	0.12	2.9
Unknown Village	147	0.83	0.20	0.35	2.3
Sargwat					
Mwahavul	21	0.03	0.67	0.05	12.54
Ganguk	37	0.05	0.54	0.09	5.71
Doemak	92	0.12	0.88	0.23	3.82
Gankogom	148	0.20	0.80	0.36	2.23
Merniang	267	0.36	0.93	0.59	1.57

Note: Sites were dropped from the analysis if both nearest neighbors were of unknown affiliation, or if one nearest neighbor was from a different village/alliance and the other was of unknown affiliation. The * indicates a category comprising several villages; these are cases where precise village affiliation was unknown. Since attraction coefficients become erratic when *n* is small, villages with 10 or fewer representatives are not shown.

[a]Number of sites within the study area (column 2).
[b]Proportion of sites with at least 1 co-villager within 2 nearest neighbors (column 3).
[c]Probability of another site's being a co-villager if locations were random with respect to village affiliation (column 4). Calculated by ([column 2] − 1) ÷ (total sites − 1).
[d]Binomial probability of at least 1 co-villager within 2 nearest neighbors (column 5). Calculated by $2p(1 - p) + p^2$, where p is column 4.
[e]Village attraction coefficient, equal to column 3 ÷ column 5.

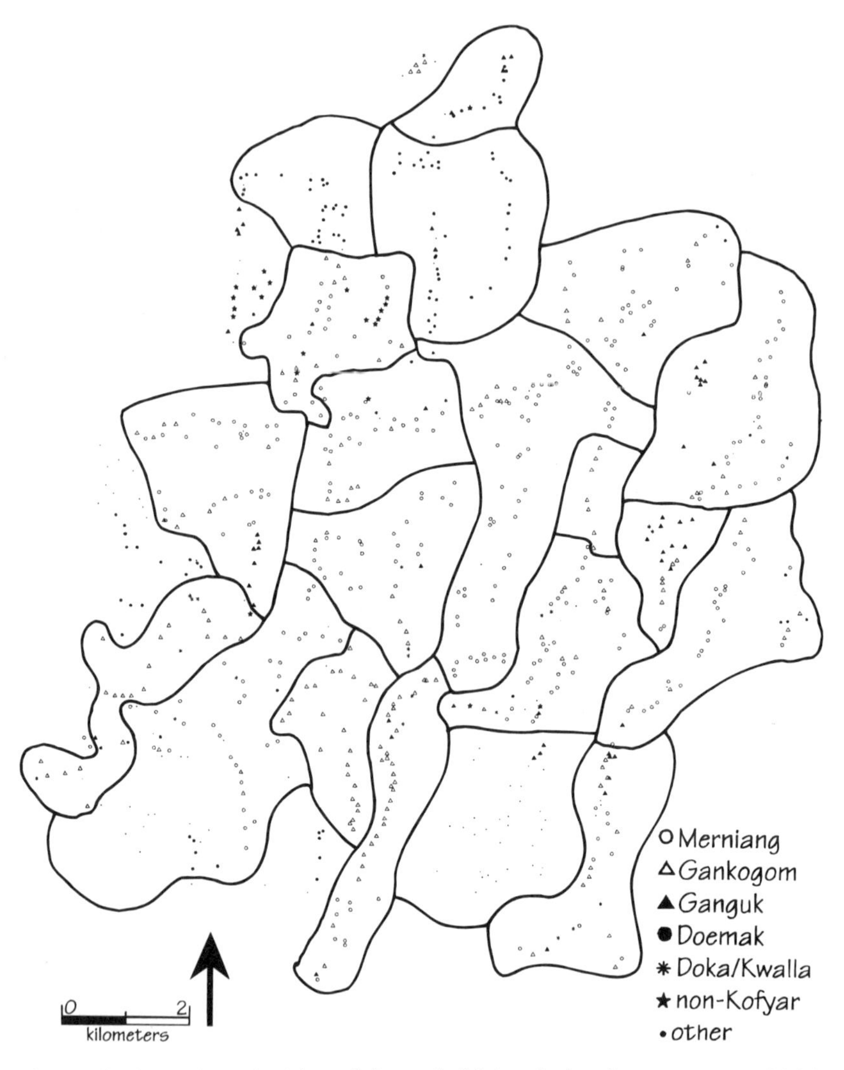

Figure 10.4. Microethnicity of household heads in the core area, 1984. (Reprinted, with permission, from Stone 1992)

from the plains communities, in which social propinquity tends to be less strong.[4]

Farmsteads are thus pulled by both the village and the sargwat affiliation of other settlements. Attraction as measured by the affiliation of nearest neighbors is generally stronger at the level of the village, especially for the hill villages, which have high levels of internal social

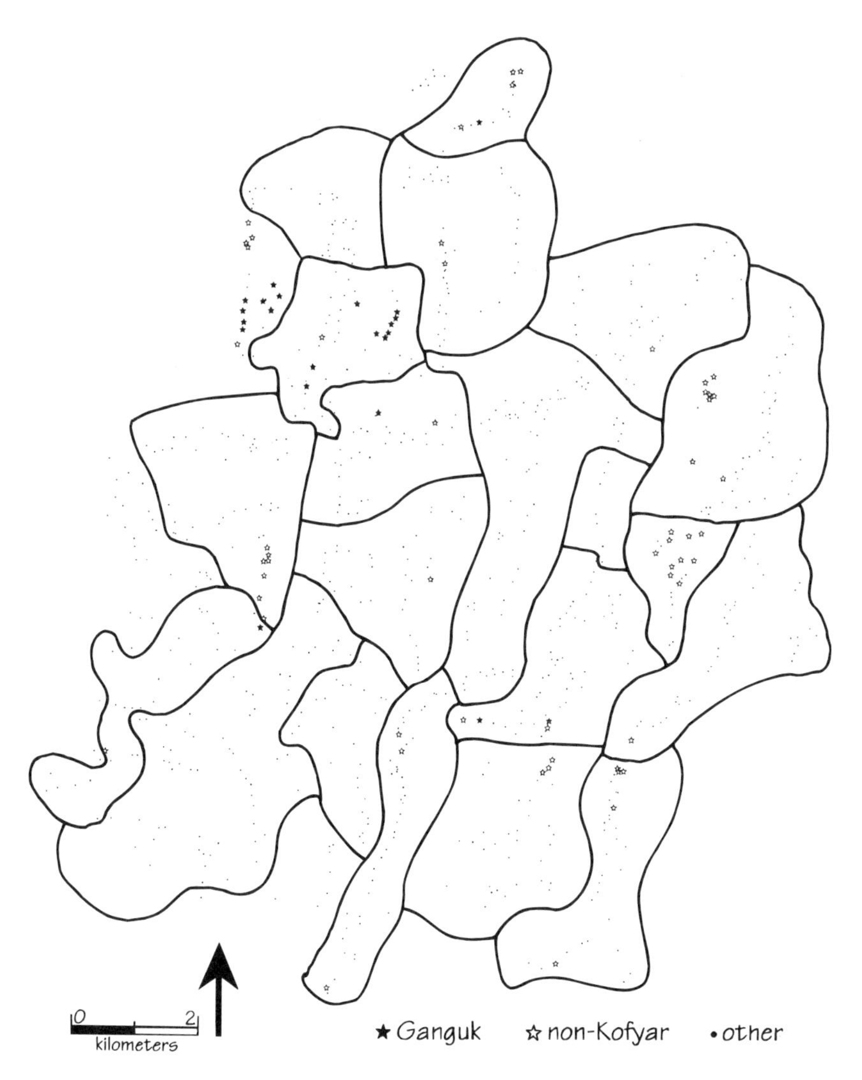

Figure 10.5. Settlement encystment in the core area, 1984. (Reprinted, with permission, from Stone 1992)

propinquity. Attraction as measured by spatial concentrations of members of particular social taxa is strongest at the sargwat level, although there are some village-level concentrations. Some large sargwat concentrations include several ungwa, and small subungwa concentrations, or encystments (fig. 10.6), occur among settlers socially distant from surrounding farmsteads.

Figure 10.6. Settlement of Mwahavul farmers in Koprume. Four of the eight Mwahavul compounds in a fairly tight cluster are visible. In appearance these compounds are typical for the core area.

As measured by presence or absence of social taxa within the entire study area, there is at least a slight attraction among neighboring inclusive tribes from the plateau but a repulsion by Tiv farms. These arrangements facilitate agricultural collaboration among the socially close, as I argue earlier in this chapter.[5]

11

Settlement and the Physical Landscape

The preceding chapters have explored relationships between settlement pattern and agriculture in the context of a social landscape. But the social process of food production is obviously acted out on a physical stage on which various landscape features directly affect settlement and other spatial aspects of behavior. This chapter deals with how settlement is affected by features on the physical landscape, with variation in water resources, soils, and other features.

In concentrating on the interplay among land pressure, agricultural intensity and settlement shifting, the theoretical framework presented in chapter 3 held constant landscape variables not directly related to the agricultural intensification slope. One such variable is water—which, because it is heavy, usually awkward to transport, and always essential—is invariably a consideration in locational decisions. The conventional wisdom in archaeology and geography is that farm settlements are attracted to water; in *Rural Settlement and Land Use,* Chisholm weighted water twice as heavily as arable land (Chisholm 1979:96). A "wet-point" site (the location of which is a function of water supplies) has long been assumed to occur where water was scarce, whereas well-watered landscapes allowed freer play of cultural factors (Johnson 1958:554). Yet exceptions turn out to be nearly everywhere, and the issue would benefit from a rethinking. In this chapter I move beyond the assumption that water exerts a constant attraction on rural settlement location, using the Kofyar data to demonstrate a marked change in the effect of water on settlement as land pressure rises.

The Changing Attraction Value of Water

The Kofyar agricultural frontier is mostly within the catchment of the Dep River, a tributary of the Benue (fig. 4.1). This catchment contains portions of the Benue Piedmont, Namu Sand Plains, and Jangwa Clay Plains physiographic zones (fig. 5.1); for hydrography it is useful to also define a transition zone comprising the southern edge of the sand plains and the northern edge of the clay plains. The drainage pattern is dendritic throughout, but the stream density varies. None of the drainages flows in the dry season, but the Dep and Shemankar usually have isolated pools of water.

The water table fluctuates seasonally with precipitation. In the Benue Piedmont and Namu Sand Plains, dry-season water is available from shallow pits dug into large streambeds. In the Namu Sand Plains there are also some walk-in wells near streams and a growing number of bucket wells. But on the Jangwa Clay Plains, there are water problems, especially as one moves south toward the Dep. In places such as Langkaku and Hayen Obi, the dry-season water table is very hard to reach with a well by January. The Jangwa Clay Plains have the reverse problem during the rains: too much water. The clays drain poorly, causing gullies that can erode and run through compounds. In some parts of the Jangwa Clay Plains, Kofyar have to sleep on raised platforms and build their fires on top of sheets of roofing metal.

The Kunamu and the Maiburugu are tributaries of the River Dep. The watershed between them runs approximately 3 km south of the Namu-Lafia road (fig. 5.1). The piedmont north of this has a medium-to-fine drainage texture and low agricultural potential. South of this, the land drops off somewhat more steeply and regularly, to an elevation of 100 m at the River Dep 20 km to the south. The drainage network generally has a medium-to-fine texture, but along its northern edge there is a high density of low-order streams. These low-order streams along the rim of the Bena (Bana) and Gbogbok drainages produce a zone within which most locations are close to water (fig. 6.1). This was one of the areas that first attracted Kofyar migrant farmers.

All of the streams in the study area are ephemeral, including the larger rivers such as the Dep and the Shemankar. For dry-season water, the Kofyar excavate into the streambeds. In the large streams, where the water table is just under the bed for much of the dry season, many small pits are dug. On smaller streams it is often necessary to dig walk-in wells, sometimes more than 5 m in depth.

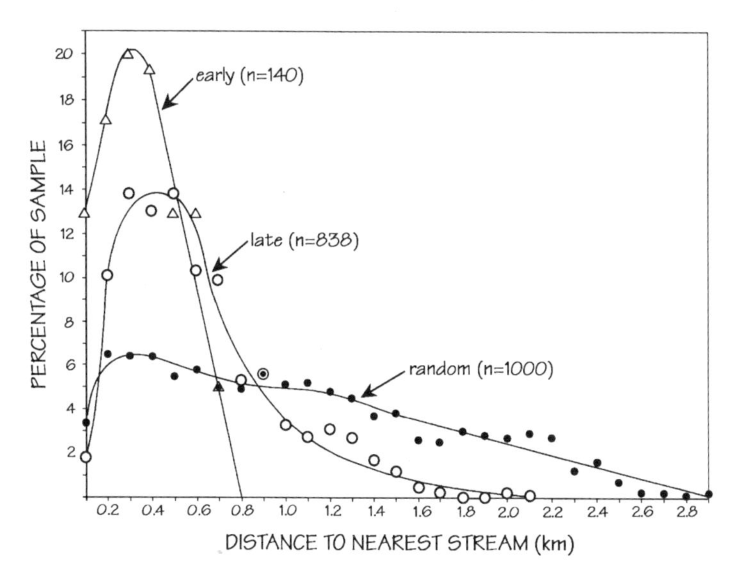

Figure 11.1. Distances to nearest water in the core area.

The location of water resources has played a key role in the evolution of settlement pattern on the frontier.[1] The initial settlement on the Namu Plains was the macrocompound of the chiefs of Kwa and Lardang, 700 m northwest of a branch of Rafin Gogo. This was the first spot along the path south of Namu that was within 700 m of a stream. When this settlement fissioned, the daughter group constructed Goewan gari 200 m northeast of the eastern branch of the same river (see chapter 6).

For each 1963 and 1984 settlement location, I calculated the distances to the closest point on a stream. To compare the distribution of distances to nearest water to the distribution that might be expected if settlement was not drawn to water, I calculated the distance to nearest water for 1000 random locations within the settled area. The distributions for the random points and the 1963 settlements (*n*=140) are compared in figure 11.1. The mean distance to the nearest stream in 1963 was 310 m (s.d.=170 m; median=300 m) as compared to a mean of 1050 m (with a median of 960 m) for the environmental background, illustrated by the random points. Figure 11.1 shows a sharp drop-off in the early distribution after 400 m, with only 5% of the compounds being located beyond 600 m and none farther than 700 m from water.

The 1984 spatial data (*n*=838; obtained from aerial imagery and ground survey) show a change in the relationship between settlement loci and streams. The mean distance to water increases to 560 m (with a median of 480 m), and the standard deviation is more than double what it is in the early distribution (350 m, as opposed to 170 m). Figure 11.1 shows that the mode has increased and the distribution is positively skewed, with more than 37% of the settlements beyond the 600 m range. There are several factors behind this development.

The first is that many of the compounds that were initially situated very close to streams had been relocated within the same farm boundaries. This was especially true in areas like Ungwa Goewan, where early compounds near Rafin Gogo were too scattered to be connected by a road; as a road took shape, these compounds were gradually pulled toward it. The settlement history of Boyi Pankok of Ungwa Goewan (chapter 6) illustrates this, beginning on the end of his farm closest to the stream and ending up across the road from the stream.

A second factor involved in the changing distribution of distances to water is the filling in of interstitial areas left unsettled during the initial colonization. These interstices do not, in general, have lower agricultural productivity, but they tend to be between watercourses (fig. 6.1).

The third factor is the expansion of settlement into areas with lower stream density. The majority of the 1963 settlements were located within the Namu Sand Plains, where there is a medium-to-fine drainage texture, and these early settlements were strongly attracted to streams. After this, settlement histories show increasing movement into the Jangwa Clay Plains with its coarser drainage texture. These developments all reflect a declining attraction value of water.

In recent years, the declining attraction value of domestic water on settlement has led to colonization of areas with even more deficient resources. This is nowhere more true than in Ungwa Hayen Obi, south of Assaikio, where Kofyar first established settlements in 1981. Although Hayen Obi farmers report high yields from the virgin or long-abandoned plots, the landscape offers almost no water at the peak of the dry season, and the population relies on water purchased from tankers that have begun to make circuits through the bush.

The situation is just as bad in Langkaku. Early in the dry season I visited a walk-in well that was the only local water supply. The cloudy water was home to a thriving colony of frogs, and there were several sets of Fulani cattle tracks around the water (although the Fulanis are fined if caught watering their cattle in the well). The well was expected to dry out before the end of February (several weeks before the onset of rains), after which the water has to be bought for ₦2 or more per 55-gallon drum. The Langkaku Kofyar have asked the chief of Namu for a "digging engine" to deepen the walk-in well, to no avail.

The declining attraction value of water in response to rising land pressure was observed by W. B. Morgan in his investigation of two areas in Iboland, southeastern Nigeria. Morgan compared a sparsely settled area having a high density of streams to a densely settled area having a shortage of water, finding what he called the "the surprising contrast of evenly dispersed settlement in a region of restricted water supply and highly localized, nucleated settlement in a region offering easier conditions" (W. B. Morgan 1955a:322). Today it is common for Ibo farmers in densely populated areas to live where Herculean efforts are required to provide water. I saw compounds in an extremely crowded area near Nsukka with a water harvesting and storage system consisting of an unlined pit in the yard, less than a meter deep, and almost a hundred large ceramic vessels sunk partly into the ground. During the rainy season, water is collected from the pit and transferred into the storage jars to provide for the dry season (Stone 1991b:23). Karmon (1966:47) reports cases from eastern Nigeria of women and children spending as many as five hours a day getting water.[2]

Water Resources and Intensification

The relationship among water, settlement location, and land pressure is an example of the implications for settlement that may be drawn from the Boserup model of population and land use. The Boserup theory states that farmers will cultivate extensively as long as there is sufficient land, because of the greater labor cost of intensive production. It is only when land pressure demands it that farmers perforce intensify—or, as I have stressed, move. Thus, whereas farmers do not necessarily minimize labor costs, they can be expected to avoid major labor expenditures whenever possible. And as Chisholm points out, travel to farm fields becomes increasingly labor-expensive as cultivation intensity increases.

There is a hydraulic corollary of this. Water is a resource that (1) is required in large quantities by every human, (2) often occurs in discontiguous locations, and (3) the transport of which often exacts a large labor investment in low-technology societies. Although the labor expenditure on water can be far less than what is often required to intensify agriculture, for populations under no pressure to intensify it may be the single most labor-expensive activity that the farmer has some control over. It is therefore entirely consistent with the framework relating agriculture and settlement (in Chapter 3) that farmers under conditions of low land pressure should farm extensively and give high priority to maintaining low residence-to-water distances.

The foregoing analysis has shown how in a frontier setting where most other factors were controlled—there was an abundance of inexpensive, well-drained, fertile land and a growing group of co-ethnics to serve as an agricultural labor pool—settlements were pulled strongly to water, always locating within 700 m of streams. However, even under mild land pressure, settlements were quick to move to greater distances from water. Under greater land pressure, as had developed in parts of the core area by the mid 1980s, many farmers were quick to accept much greater distances to water and to colonize areas with real restrictions on dry-season water. To the extent that the key factor in intensification is labor, such movement is a form of locational intensification: farmers respond to land pressure not by remaining and investing greater labor per unit of land in cultivation, but by moving to locations that demand greater labor in water transport.

Finally, let us consider the role of gender in the changing attraction value of water in settlement decisions. Hornby and Jones (1991:11) suggest that "the apparently low priority given to being near a water supply in many small African settlements may also reflect the fact that the decisions are made in a largely male-dominated society where it is usually the women who suffer the daily toil of carrying water to the home." With the Kofyar, it is true that water is fetched mostly by women while the scouting and acquiring of land is handled by men. Under rising land pressure, settlement may be less attracted to water because it is women's workloads that suffer.

Still, the ability of Kofyar household heads to make decisions that eat into women's work schedules is ultimately limited by the trump card held by each woman: the prerogative of divorce, which women can and do exercise to protect their economic interests (M. P. Stone 1988a). The value of this card is reflected in the very high percentages of frontier women between the ages of 21 and 50 who farm independently (M. P. Stone 1988b:295) despite the tightening land base.

Settlement and Soils

In chapter 3 I argued that Boserup's theory of intensification works as a high-level generalization but that it obscures local variability that is essential to agrospatial decision making. Locations differ in their agricultural start-up costs (and moving to lands with increasingly high start-up costs is locational intensification); they also differ in how much, and at what cost, their production concentration can be boosted by human intervention ("soil sensitivity"). Start-up costs and soil sensitivity are related; but they can also be independent, producing widely varying intensification slopes. Their spread onto the Plains of Muri has brought the Kofyar to physiographic zones with definite differences in intensification slopes. If, as argued in chapter 3, agricultural agrospatial decision making responds to local intensification slopes, we should expect patterned differences in these physiographic zones. When confronting a steep drop-off in the intensification slope, agrospatial alternatives include relocation to avoid intensification, movement into areas with increasingly high start-up costs (locational intensification), and cultivating subsidiary locations (generally by satellite settlement). Each of these strategies was used by the Kofyar.

We can explore how the role of soil variability in locational criteria has changed through time by looking at patterns of both frontier colonization and abandonment—where farmers moved into and out of. My analysis of the history of frontier colonization is based on site choices in the settlement histories. A site choice is a settlement episode in which the respondent becomes head of a farm where he was not head before. This may be either when a respondent who was already a household head moved to a different farm, or when the respondent inherited the farm where he resided and elected to remain there. It is relatively common for a man to assume household headship (usually upon the death of his father) and move to a new location as soon as arrangements can be made (usually within a year); this is counted as a site choice in the new location but not the old. Cases where a man inherits a secondary farm but does not move there are not counted as site choices. Abandonment is when the respondent moves (that is, makes a site choice) to a different ungwa without maintaining the first residence. Settlement episodes not ending in abandonment are given an end date of 1984.[3]

A detailed analysis of the colonization patterns is presented in G. D. Stone 1988, but I will describe the central findings here. In the late 1940s and early 1950s, Kofyar were moving into scattered locations in the Benue Piedmont (see chapter 5). Throughout the rest of the 1950s, as the migration rate increased dramatically, the migratory stream flowed into two principal areas in the Namu Sand Plains: south of Namu and south of Kwande, especially in the area of Mangkogom. It is significant that these two parts of the sand plains differed in soil quality, with the southern Namu bush offering optimal conditions as compared to poorer, lateritic soils south of Kwande. Initial site choices were guided by a relatively coarse soil taxonomy, consistent with our earlier conclusion that "most folk classifications of soils . . . appear to encode only basic information about soil fertility and potential crop productivity" (Wilshusen and Stone 1990:109). But however similar their early settlement rates may have been, we will see in the next section that the abandonment rates in the Namu and Kwande areas differed sharply.

Beginning in the mid 1960s, the Jangwa Clay Plains began to attract much of the new settlement, and throughout the late 1960s and 1970s, the Benue Piedmont and Jangwa Clay Plains dominated site choices. Some of the farmers moving onto the clay soils were relatively late arrivals on the frontier, unable to find space in the sand zone; others had farms in the sand zone where yields were beginning to drop; others were young families fissioning from farms in the sand zone to start on their own (G. D. Stone 1988:203). This movement into the shale-derived soils was especially pronounced in the 1970s, although it was beginning to abate in the early 1980s.

By the late 1970s, prime areas such as Ungwa Long begin to be divided, and site choices rose again. Movement into the southern clay zone dropped off sharply, and movement began in earnest onto the western reaches of the clay zone, south of Sabon Gida and Assaikio, which was the most obvious advancing edge of the Kofyar settlement frontier when my coworkers and I were conducting fieldwork.

Abandonment Patterns and Localized Intensification Slopes

Although this history is instructive as to how the colonization process is affected by soil variability, by focusing on farm acquisitions it does not isolate the role of soils in locational decisions as land pressure rises (Norling 1960). A fuller picture emerges when the site choice analysis is coupled with an analysis of occupation spans. I have done this by plotting year of site choice against abandonment year for a sample of ungwa representing different soil types. Year of site choice, as defined above, is the beginning year of a settlement episode in which the respondent becomes head of a farm, and abandonment is the year in which the respondent moved out of the ungwa.[4] The year of abandonment minus the year of site choice gives the ungwa occupation span of the settlement. A cross-tabulation of ungwa occupation span by year of site

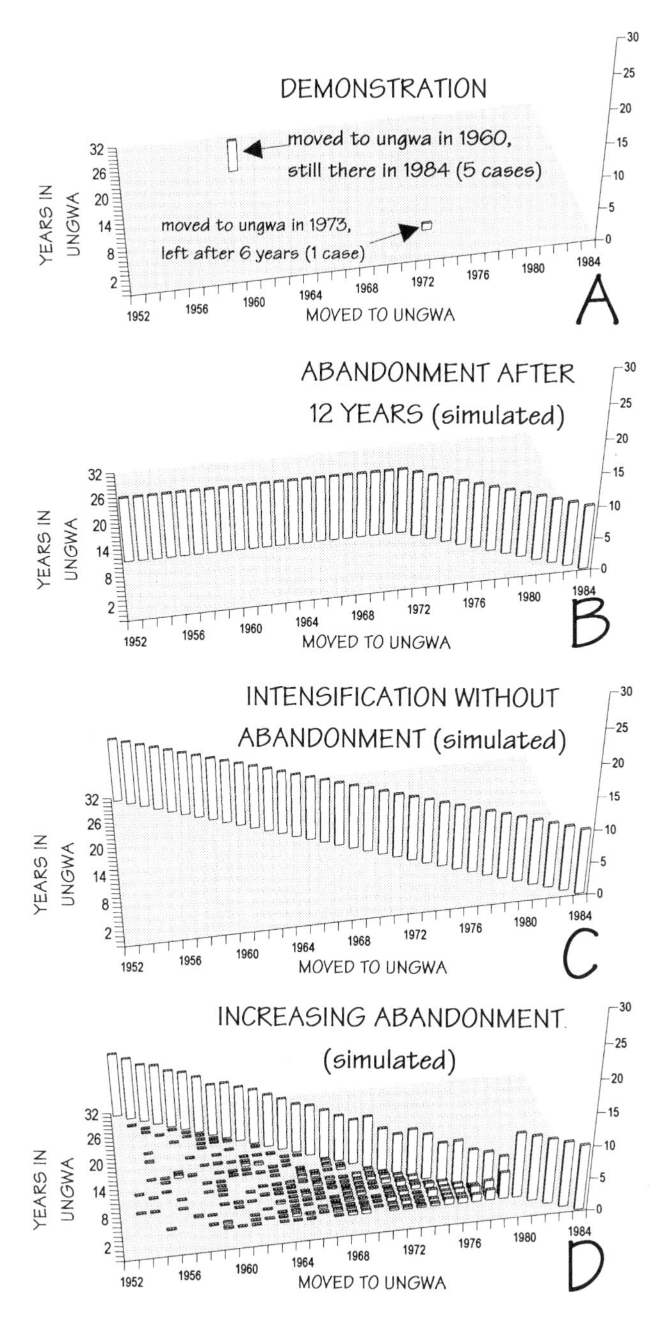

Figure 11.2. Abandonment profiles.

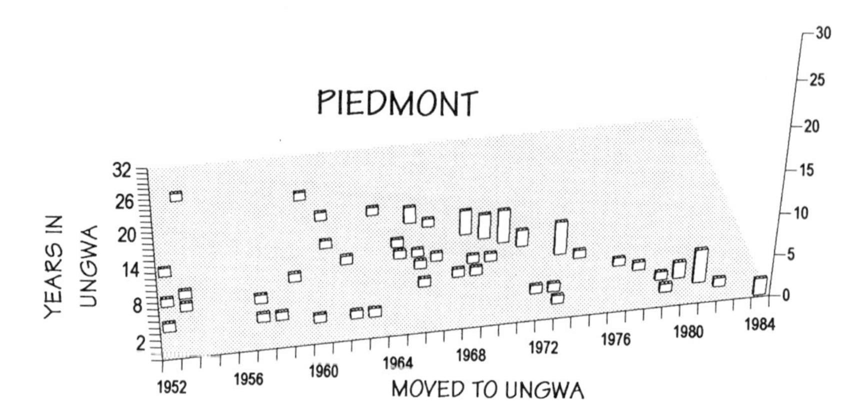

Figure 11.3. Abandonment profile, piedmont zone.

choice gives the abandonment profile for particular zones. The accompanying figures are graphic representations of these abandonment profiles.

Figure 11.2a shows how the settlement histories are being used, with examples of theoretical patterns resulting from particular responses to the intensify/abandon option. Figures 11.2b–11.2d all assume a constant influx of 10 settlers per year (z-axis). Figure 11.2b is the hypothetical case of shifting settlement in which all settlers abandon the area after 12 years. This is an idealized representation of the pattern characterizing most of the cases of shifting cultivators discussed in chapter 3. Figure 11.2c is the idealized representation produced when no farmers abandon the area—all elect to stay and intensify production, as in the Boserup model. Figure 11.2d shows the sort of pattern resulting from early colonists electing to remain but later arrivals increasingly leaving in response to declining agricultural yields. This pattern is less neat but more realistic than the first two patterns. It may arise, for instance, when the first settlers in an area claim the best or largest plots (Eidt 1976:9). The abandonment pattern in figure 11.2d was simulated using these rules: (1) 10 new farmers are added each year from 1952 to 1984; (2) farmers do not consider leaving until they have farmed for a minimum of 5 years; (3) the probability of leaving climbs from 0 for those arriving in 1952 to 0.5 for those arriving in 1979. (Those arriving after 1979 will not have been there long enough yet to consider abandonment.)

Figure 11.3 shows the actual abandonment pattern in the 64 settlement episodes in the piedmont zone. Since the late 1940s, Kofyar have

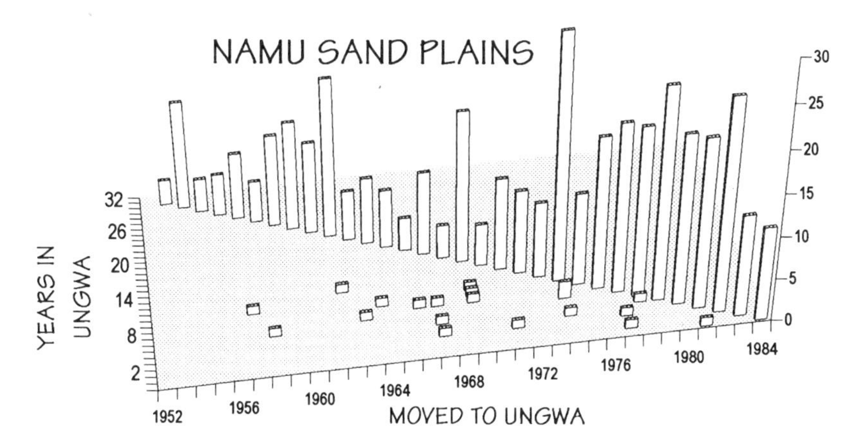

Figure 11.4. Abandonment profile, Namu Sand Plains.

consistently left piedmont farms, usually after 4–12 years; only a scattering of farms have been kept for more than 15 years. The pattern in figure 11.3 is of course not as neat as the hypothetical example of consistent abandonment (fig. 11.2b), but it is one of the better real-life examples of shifting settlement. All of the settlements before 1960 and 83% of the settlements before 1968 have been abandoned, usually after 3–12 years.

This sharply contrasts with the abandonment profile in figure 11.4, based on the 394 settlement episodes in the sand zone south of Namu. Despite the history of consistent in-migration since the early 1950s, abandonments in the sand zone are quite rare. There are a few scattered abandonments of farms, but for most periods the abandonment rate is below 10%, and the overall abandonment of headed farms is just 5%.[5] Given the many circumstances that can affect farmers' decisions, it is unlikely that such a large sample of settlement histories would ever produce the perfect pattern of nonabandonment (fig. 11.2c), but the sand zone certainly comes close.[6]

However, in the clay zone there has been a gradual transition from settlement stability to a marked tendency for abandonment after only a few years. The first ungwa in the Jangwa Clay Plains to be settled was Ungwa Kofyar, followed by the southern end of KDG Koegoen and Duwe South; the soil drainage problems were least severe here, and there have been no abandonments (table 11.1). But the 1970s saw settlers pouring into the less desirable parts of these ungwa and into newly opened ungwa to the south, where drainage problems were wide-

Table 11.1 Farm Abandonment Rates by Soil Zone

	Total Episodes	Abandonments
Sand Zone		
Wunze	62	0
Goewan	47	3 (6.4%)
KDG Zang	43	2 (4.7%)
Hanyar Kwari	41	5 (12.2%)*
Long	35	0
Dunglong	29	2 (6.9%)
Kwallala, Goejak	29	1 (3.4%)
Duwe North	28	1 (3.6%)
Koprume-Dayim	26	0
Koprume-Dakup	12	0
KDG Koegoen—sand	12	0
Rafin Gwaska—sand	10	0
Koprume-other	9	0
Pangkurum	7	0
Kopdogo	6	1 (16.7%)
Gogot	6	1 (16.7%)
Koedoegoer Raplong	5	0
Bubuak (non-Bala)	3	1 (33.3%)
Other	5	3 (60.0%)
Southern Clay Zone		
Kofyar	45	0
Dangka	40	12 (30.0%)
Duwe South	31	0
Rafin Gwaska—clay	29	0
Langkaku	24	10 (41.7%)
Kangiwa	24	4 (16.7%)
Kwari	15	12 (80.0%)
KDG Koegoen—clay	12	0
Burugu	1	1 (100.0%)
Western Clay Zone		
Hayen Akuni	12	0
S/G Dawam	9	0
Hayen Obi	8	0
Assaikio	2	0

Piedmont Zone		
Naburuk	26	3 (11.5%)
Njak Area	18	17 (94.4%)
Mandeshik	10	1 (10.0%)
Shendam Area	15	14 (93.3%)
Kwande Zone		
Mangkogom	26	2 (7.7%)
Bakin Ciawa	4	4 (100.0%)

*I suspect this figure is inflated by inclusion of cases from the Kwari area.

spread. Occupation spans began to decrease, becoming especially short in late-settled ungwa well south of the core area. I visited one ungwa near Langkaku in 1984 where only a handful of farmers—those on the largest and choicest plots—were not looking for new locations.

Figure 11.5a shows the abandonment pattern for the Jangwa Clay Plains. It does not closely resemble the simulation of increasing abandonment (fig. 11.2d) because immigration into this zone has been considerably different from the steady influx assumed in the simulation. As noted earlier in this chapter, the arrival rate on the clay plains was very low until the mid 1960s, and then high until it began to taper off in the early 1980s. If we run the simulation of increasing abandonment using the same rules as in figure 11.2d but inserting the actual number of arrivals for each year, the result is the abandonment pattern in figure 11.5b. It is strikingly similar to the actual data, and I believe the rules for this simulation are a fair approximation of the changing abandonment strategies on the Jangwa Clay Plains.

The abandonment profile of the southern Kwande bush (mainly Mangkogom and Bakin Ciawa) is shown in figure 11.6. The initial rush of settlers, coincident with the first colonization of Ungwa Long, shows up at the peak in the early 1950s. Since then, there has been a low but fairly consistent rate of site choices, with a moderately high percentage of abandonments.[7]

The data on settlement trends and abandonment profiles show that the criteria for site choices are different from those for abandonment/intensification decisions: the population that poured simultaneously into Ungwa Long and Mangkogom reacted very differently as land pressure rose. Much of the patterning in abandonment rates is explained by differences in local impediments to production concentration, and thus

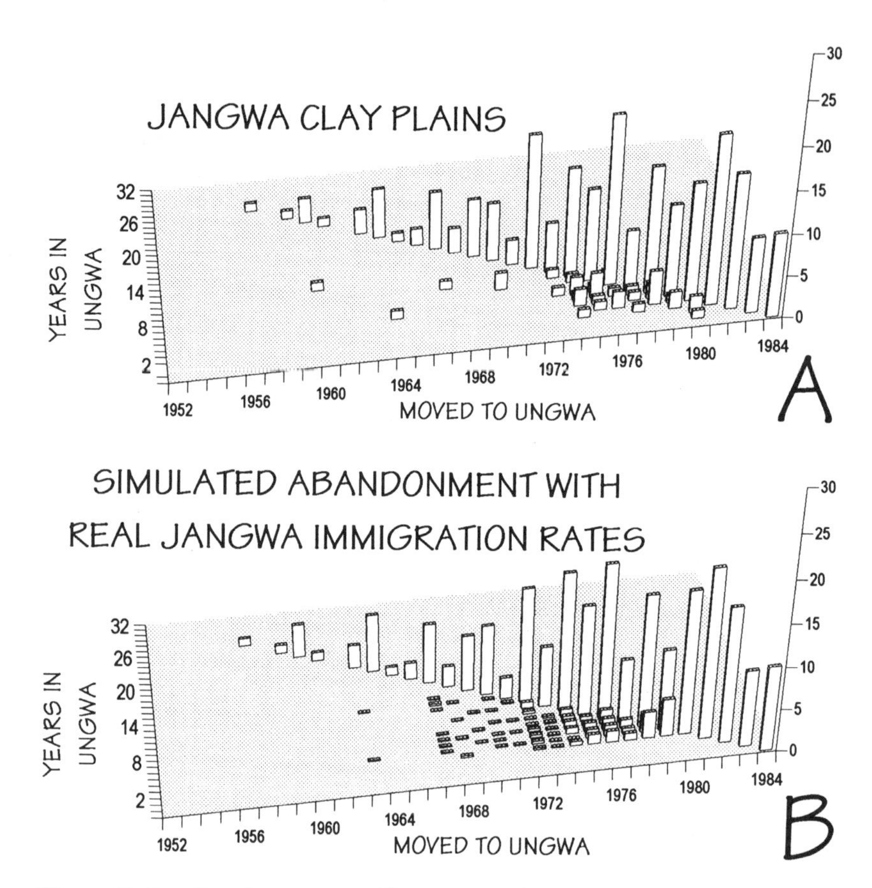

Figure 11.5. Abandonment profile, Jangwa Clay Plains.

different intensification slopes. Farms in the sand zone (the optimal part of the Namu Sand Plains) experience a very gradual drop-off in fertility and invasion by weeds including the parasitic *striga*. The soils offer high sensitivity and high marginal returns to intensive farm labor. Flagging fertility can be mitigated by light applications of dung or chemical fertilizer, and the agricultural work calendar can be rearranged by making next year's yam heaps in this year's sorghum field, which levels the planting bottleneck and kills the striga while performing routine weeding. Farmers on the Namu Sand Plains have the highest rate of dung fertilizer (*zuk*) use (43% of the sample of 571 farms), reflecting the pattern of intensification rather than abandonment in this zone.

The incidence of zuk use (25% of the sample of 24 farms) on the gravel soils of Mangkogom is somewhat lower, showing that the ratio of

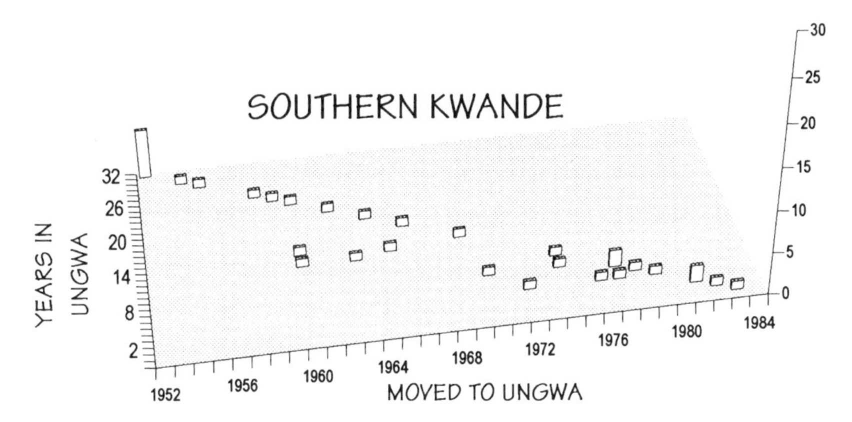

Figure 11.6. Abandonment profile, southern Kwande area (mainly Mangkogom and Bakin Ciawa).

abandonment to intensification is higher than on the superior land south of Namu. Land use is interesting in Mangkogom. It is the land freed up by the many abandonments in the 1970s that has allowed many of the earliest settlers to remain. Population density in Mangkogom is approximately 60/km^2, and my land-use survey showed only half of the land to be in crops. Rather than fertilizing their own plots, residents shift cultivation among abandoned and rejuvenated plots—in effect, extensifying agriculture in response to dropping land pressure. Farmers here often adopt plots several hundred meters away from their residences.

The clay soils too have seen abandonment over intensification, but their ecology has been different. Rather than Mangkogom's laterite and the sand zone's fertility and weeds, the problems here have more to do with drainage. Some of the farmers in Jangwa Clay Plains areas such as Dangka and Kangiwa found that after a few years of cultivation, drainage on the clayey soils was often seriously impeded. In some areas, the greater surface runoff meant that in dry years, yams may receive insufficient moisture and may rot in the heap. The reduced infiltration after cultivation likely results from increases in soil bulk density (Hecht 1981:94).[8] The problem varied locally, with many farms at relatively high locations offering good drainage; these tended to be selected by early-arriving pioneers and therefore tended to be especially large as well. But for the smaller, poorly drained plots, the intensification slope drops off very sharply; I do not know how additional field efforts could solve the drainage problem, and neither do the Kofyar I talked with. Instead, they were increasingly resorting to abandonment, as described earlier in this chapter. The low incidence of zuk use here (only 10% of the sample of 181 farms) reflects the reluctance to intensify cultivation on clay soils.

The piedmont received the first migrant farmers in the 1940s and a later influx in the 1960s. Most of the early pioneers practiced shifting settlement, maintaining their farms in the homeland. Abandonments were precipitated by land pressure—not rising population density but degradation of productive resources. But by 1984, Kofyar response to declining yields was mixed. Farmers moving to the piedmont in the mid 1960s found the movement options fairly discouraging by late 1970s. Land was tight on the Namu Sand Plains, and the piedmont farmers, who were mostly hill farmers from the Ganguk area, had few connections in the sand ungwa. The southern clay zone was seen by most as a flop, and the frontier had moved to the relatively distant western clay zone.

Lacking the sort of obvious or necessary response, like intensification in the sand zone or abandonment in the southern clay zone, the piedmont farmers have divided their responses. Some have left for recently started ungwa on the western clay zone; there is a new enclave of approximately a dozen Ganguk farmers in Hayen Obi, all of whom have come from the piedmont. These are mainly young families, who tend to be more mobile than the older generations. Others have stayed and intensified, at least on the smooth piedmont. The impediments to production here are weeds and infertility, but the soils are poorer to begin with. Nevertheless, the sample shows that 39% of the 28 farmers on the smooth piedmont were manuring their fields, as contrasted with none of the 13 farmers on the rocky piedmont.

Soils and Settlement in Context

The Kofyar data show soil variability to play a vital, but not simple, role in shaping settlement pattern. The soil criteria for site choices avoid the worst soils but place little premium on optimizing soil quality. The criteria for intensifying, rather than abandoning, under land pressure are more stringent, linked directly to the marginal returns reflected in the localized intensification slope. The role of soils in settlement decisions is roughly the opposite of that of water, which strongly constrains settlement until land pressure rises, when its influence begins to diminish.

Morgan's (1955a:322) study also showed compounds in the crowded area to be strongly attracted to the best agricultural lands, compared to compounds in the sparse area, which were attracted to water. Also comparing land use on different soils in Nigeria, Gleave and White (1969:284) found dense population to produce "a much closer adjustment to the minutiae of physical conditions" in crowded areas. Paralleling this is Padoch's (1986:283–288) finding that in selecting plots for intensive irrigated pond fields in Indonesia, minor differences in topography and water become more important than in selecting land for swidden agriculture (see also Bogucki's 1987 study of agrarian site locations on the northern European plain).

These patterns underscore the need to move beyond the Boserup model of intensification's not adapting to soil quality to an understanding of agrarian settlement decisions based on localized intensification slopes. When abandonment is impossible, land pressure may indeed produce pressure to intensify production across different soil and environment types. Yet when farmers have the option of abandoning degraded land, soil quality may play a key role in the intensify/abandon decision.

Intensify/abandon decisions are made on the basis of agrospatial alternatives as well as the marginal return to intensifying locally. The range of alternatives is determined not simply by population on the frontier, but by the farmer's knowledge of and ability to acquire available land. Land access is largely a matter of social networks, as shown in Berry's (1993) comparative study. In addition, as I showed in chapter 10, networks based on social propinquity in the homeland were instrumental not only in facilitating movement into new areas, but in encouraging immigrants to join pioneer communities.

Markets and Agrarian Settlement Decisions

Bronson (1975:215) writes, "A group's desire to remain near any resource—a river for the sake of communications, a spring for drinking water, a good flint mine . . . or perhaps a city—could cause the group to intensify rather than to move to a less crowded place. Variations in the intensity of land use can thus be caused by factors that have little to do with variations in the land's intrinsic quality." No features of this type on the nonecological landscape are of as much concern to the Kofyar as markets.

The role of central functions in shaping settlement patterns lies at the heart of Christaller's central place theory (1966). We have been looking mainly at relationships between settlement pattern and the practice of agriculture, but Kofyar making settlement decisions certainly have not been oblivious to the towns scattered across the Namu Plains. Especially when talking with young adults, it is clear that the small but bustling town of Namu was a real attraction over the slower-paced home communities. Yet the locations of agrarian settlements have consistently shown only a very weak pull toward towns and the central functions they offer.

I believe the first reason is that, however important the town's goods and services, they do not have to be accessed frequently enough for settlements to be attracted by the proximity-access principle. Although the Kofyar have seen impressive increases in their disposable income, with the average household taking in ₦1160 from crop sales in 1983–84, a substantial proportion of this is saved for big-ticket items like bride price, housing, or a motorcycle. Indeed, many Kofyar look down on frittering money away on small luxuries, like restaurant meals with meat.[9] Many Kofyar value access to small clinics offering "Western" medical treatments such as pills and injections, but the access rate appears to be too low to have much impact on settlement location.

Most of the farmers' trips to town are for the periodic markets. Part of the reason is to sell small, portable amounts of produce. But a prime motivation for the trip to market is socializing, or *soe loetuk*, which literally translates to "eating the market," but which could be said to mean "working the crowd." Towns and their periodic markets allow this to happen on a large scale, but it happens on the countryside anyway; mar muos often turn into fairly large parties, and during the dry season hardly a day passes without there being a *pe muos*—a "beer place," where people congregate.

Further reducing the attraction of Kofyar farmsteads to central places is the increasing spread of secondary (nonagricultural) functions on the countryside. I became intrigued with the notion of noncentralized central functions after a trip to the farmland surrounding Nsukka, an area of exceedingly dense agricultural settlement in the Ibo heartland in south-central Nigeria. There one finds mile after mile of contiguous farmsteads practicing highly intensive agriculture. Quite a number of the farms also offer secondary functions; they sell food or palm wine, bush meat, medicine, clothes, and a multitude of other wares. They seem, as M. J. Mortimore has noted, to be "urbanizing in situ" (personal communication, 1984).

The same process is beginning in Namu District. By 1984, several ungwa were beginning to hold small periodic markets. Like the periodic markets held in all towns in the area, these ungwa markets were held on a designated day of the week and in a designated location. But whereas town markets might have several hundred vendors proffering a dizzying variety of merchandise and might meet year-round, ungwa markets offer only a few goods or services and are mainly restricted to the dry season. Most are located along roads with a lot of foot traffic. They are usually marked by an isolated ramada, which falls into disrepair during the agricultural season.

The diminutive dry-season market near the compound in Dadin Kowa in which my coworkers and I lived consisted of a tailor with an old foot-pump sewing machine; a sundries merchant from Namu with batteries, candies, and kola nuts; and a few girls with millet beer for sale by the jug or by the calabash-ful. A little market was also starting up in Ungwa Long beside the dirt road to Namu. It was to be held on Namu's market day, to cater to the considerable foot, bicycle, and taxi traffic.

Diverse secondary functions were appearing in individual compounds as well. Several farmers acted as beer retailers in the dry season, importing caseloads of bottled beer to be resold at a profit in the bush. One man even constructed a saloon by his compound, where he sold bottled beer from Namu. This unshaded, unwindowed, aluminum box had all the ambiance of a pizza oven, and it occurred to me that in another cultural context, he could have called it a dry sauna and charged just to sit in it. But Kofyar social life traditionally revolved around the pe muos, and I am quite sure this miniature saloon, as a perpetual pe muos, turned a profit. Other secondary functions included gasoline-powered grinding mills and churches.

The attraction of the residence to town is mitigated by reliance on wheeled transport. Bicycles and motorcycles (and in a few cases, cars and trucks) have been some of the most common capital purchases. Of 3327 adults on frontier farms, 13.8% reported owning bicycles and 9.7% reported owning motorcycles, whereas 0.3% owned a car or pickup. Vehicle purchases have not been uniform in space; adults owning a vehicle are more than 40% in ungwas such as Kwallala, KDG Koegoen, Ungwa Kofyar, and Duwe South (which are more than 9 km from Namu) but less than 30% in almost all closer ungwas.

All of this suggests that intensification's pull of the farmsteads toward the plots should be a much more potent influence on settlement location than the attraction to market towns and their secondary functions.

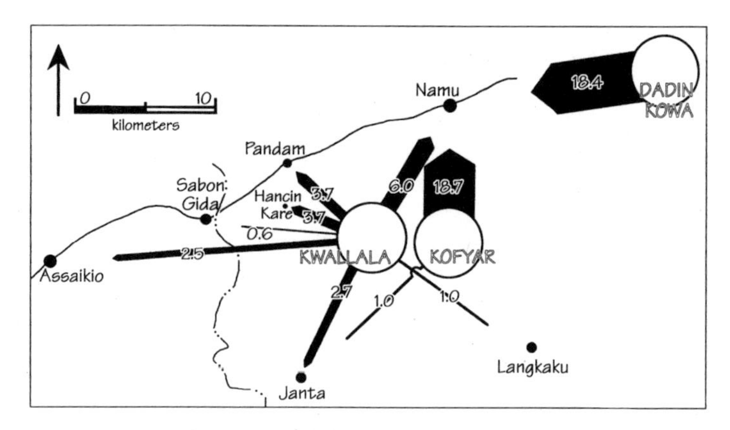

Figure 11.7. Rates of travel to local markets. Bar widths and numbers indicate mean number of trips per person per year to the indicated locations. Destinations with travel rates less than 0.5 trips/person/year are not depicted. Data are from the labor sample households.

Figure 11.7 shows the access frequencies of our 15 labor study households to towns in the area. Bar widths indicate frequency of travel; for instance, the average person in Kwallala made 6.0 trips to Namu and 2.5 trips to Assaikio during the 1984–85 agricultural season. The great majority of trips were for the local periodic markets. The pattern is generally more consistent with farmers attending whichever markets happen to be nearest than with farmers being pulled to towns for the goods and services available there.

Roads

Throughout its development, the core area settlement pattern has had a distinctly linear character. This was obvious in the 1963 map (fig. 6.1) and was still clearly discernable by 1984, even after the proliferation of side roads and interstitial paths. The linear quality is one of the most apparent contrasts between the homeland and frontier settlement patterns. The difference is partly due to the contrast in production of surplus for market. The evacuation of millet and yams to the Namu market requires mechanized transport and roads. Kofyar evacuate their crops both by selling to traders who visit the farms in lorries and by hiring small pickups to transport produce to local markets. Growers also bring their harvest piecemeal, sometimes by headloading but more often by bicycle or motorcycle.

The Kofyar build and maintain many of their own roads and motorcycle bridges. These are used for trucking crops out in the dry season and otherwise handle foot, bicycle, and motorcycle traffic. Roads are carefully maintained by communal work parties. Some feature erosion-control ditches and speed bumps that keep water from flowing down the road.

With frontier road building also came a new relationship between roads and compounds. In the hills, compound locations were determined by topography and by the locations of other farms. Numerous footpaths connected compounds to each other and with outfield plots. There were no roads; roads in the settlement belt below the escarpment had little effect on compound siting.

But on the frontier, there was from the beginning a marked pull of compounds to roads. The relatively flat landscape posed few constraints to settlement placement; by arranging their compounds in close proximity along a road or path, with farms extending in both directions perpendicularly, farmers could both be near the center of their farms and also move easily between farms, which was important for group-labor mobilization.[10]

Thus, the change from paths adapting to compound loci to compound loci adapting to roads is due to changes in topography, the need to evacuate agricultural surplus, and the pattern of labor mobilization. This effect of surplus production has been seen elsewhere in Nigeria, as in Udo's (1961) discussion of cash-crop farmers in Otoro District, where "ribbon villages" develop, reversing the subsistence-economy phenomenon of paths adapting to compound locations (Udo 1966).

One indicator of the attraction to roads is the way that residences are abandoned and rebuilt in response to shifts in road location. The hamlet of Dadin Kowa, for instance, which had been clustered along the old dirt road connecting Namu and Kwande, was in large part rebuilt a few hundred meters to the south when a paved road was built there in 1980.

In contrast, a recent twist on the homeland pattern of roads adapting to residential locations is the construction of roads to hill villages. Remarkably, by 1985 the Kofyar had nearly finished carving a dirt road up through Lardang to the high village of Bong. Inaccessibility makes hill villages vulnerable to abandonment, and the Kofyar are keenly interested in keeping this from happening. Bong residents pointed out with pride that this road would allow them to grow cash crops at home, by allowing the evacuation of crops that could never be carried in sufficient quantities on foot. They were also looking forward to having bags of chemical fertilizer and building materials delivered by truck. In a larger sense, the road is intended to "fix" the settlement and to prevent abandonment. Because maps of these rural areas tend to focus on roads and the communities connected by them, this road will literally put Bong on the map.

12

Agrarian Ecology and Culture

This study of the determinants of an agrarian settlement pattern began with the notion of settlement systems as sets of rules. Although I have pointed out problems with that approach, I have not discarded it, because it has undeniable heuristic value in dealing with what can be a very complex subject. Moreover, as Schiffer has argued to archaeologists, explanations are generally constituted by rules or lawlike premises, even if the rules are implicit and the laws are only statistical (Schiffer 1976).[1] Therefore I return to the notion of settlement systems as sets of rules, but I can now strengthen this approach in two fundamental ways.

The first has to do with the nexus of human activities in which locational rules are embedded. I have argued that rules of agrarian settlement are embedded in the ecology of agricultural production, and I have explicitly linked agrarian settlement with intensification theory, which has been curiously neglected in the settlement literature. This strengthens our understanding of intensification theory and settlement theory alike. For instance, an obvious distillate of Boserup's work is the "rule" that rising rural population density necessarily causes agricultural intensification. Although this rule is still a highly influential statement on human ecology, three decades of interdisciplinary research have downgraded it from a revelation to a near-oversimplification. Many of the limitations are directly related to settlement, and by better understanding relationships such as that between intensification and abandonment, we better understand intensification itself.

The second has to do with the determinism inherent in settlement rules. Flannery's suggested rules are unequivocal and unconditional:

"[E]xpansion will take the form of daughter communities along the river, midway between the original community and the limits of the valley" (1976a:180). But, aware that no set of sites will completely conform to a single set of rules, Flannery states the rules to be neither jural nor deterministic, but rather probabilistic: they are acting on processes we must think of as random (Flannery 1976b:171). This seems to dichotomize all variability into either (1) patterning explained by a rule or (2) noise. But it was clear to me from the outset that Kofyar settlement decisions were mediated by numerous sets of priorities, more like simultaneous equations than a linear set of rules. The key was that the different priorities favored different locational solutions, and there was variation in the value of following (or cost of neglecting) each priority. The priorities and their relative values also changed through time. All this means that we must consider the "strength" of rules distilled from an agrarian settlement system: we must ask not only what locational solutions are favored and why, but how other solutions may take precedence.

Given this framework, I will review my explanations of the key aspects of the development of the Kofyar settlement pattern in terms that at least approach the notion of rules. I will then turn to the more philosophical issue of why these rules, and not others, should have been followed.

The Rules of Settlement

Intensification Theory and Dispersion

I began with my own version of intensification theory, in which the driving force is population pressure, or a rising ratio of food demand to the quantity and quality of productive land. Land pressure does not automatically lead to intensification *or* abandonment, but to a choice between the two. Other factors being equal, patterns of abandonment versus intensification will respond to the localized intensification slope, or the effect of marginal work in agriculture on labor efficiency.

Intensification may follow various trajectories, but paleotechnic intensification often requires increasingly frequent labor inputs, which exerts a pull on residences because of the proximity-access principle. The premium on residing on or near the plot is proportional to the frequency of plot access. When inputs are divided among more than one location, this pull is diluted; the more concentrated the inputs are in space, the greater the premium on dispersion. When this pull is strong, it can override strong incentives for agglomerated settlement away from the farmland. When the pull is weak, it may be overridden by even minor defensive considerations, such as the thievery by baboons that Kofyar pioneers complained of, or by convenience for religious rituals.

Intensification and the Deployment of Laborers

Whereas most arguments in intensification theory have hinged on overall labor and marginal returns, it is labor scheduling that has been most instrumental in the shaping of Kofyar settlement patterns.

The Kofyar have mobilized labor to meet the intricate demands of intensive production by various mechanisms, including a collaborative system that capitalizes on the economic and ecologic advantages of millet cultivation. The practice of regularly working on others' fields means that residences are pulled not only to their own plots but also to other residences (settlement gravity) by the proximity-access principle. When the number of agricultural trips to one's own plot and to neighbors' plots is as high as it is with the Kofyar, there is a premium on farm shape and settlement spacing approaching optimal configuration for the two types of movement.

Although I have not attempted to predict the occurrence of labor pooling across dispersed settlements, I have examined reasons why it occurs, and I have argued that it is more common, and more closely linked to a range of social and spatial phenomena, than is generally appreciated. The question of how it affects settlement arrangement in other systems needs to be investigated.

Agricultural Labor and the Formatted Landscape

Collaborative labor is an important factor in the social division of the landscape into ungwa, the named areas that act as forums for labor mobilization. These sociosettlement entities have only minor and intermittent political and cadastral functions but a daily role in shaping agricultural movement and ensuring dependability in the social machinery for mobilizing labor. I suggest that as the regular pooling of labor across multiple settlements increases (as compared to neighborly help in extraordinary circumstances), so does the importance of formatting the landscape into sociosettlement units that can manage localized labor exchange. However, availability of expedited transportation would allow labor pools to be less localized and would thus diminish the need for this sort of formatting.

Agriculture and Ethnicity

I have also argued that it is labor pooling and the attendant demand for close and predictable cooperation that militates for social propinquity among ungwa members. Whereas the broad patterns of migration have largely been artifacts of local history, the fine-grained patterning in who lives near whom is closely linked to the social organization of agricultural labor. Especially interesting is the pattern of settlement encystment where a considerable cultural distance separates settlers.

This socioecological landscape has of course developed on a physical landscape that has been a prime consideration in locational decision making.

Location with Respect to Water

Water was a prime criterion of settlement location in the pioneering phase on the frontier, but neither its specific effect on settlement location nor the reason for this effect is obvious. Pioneers had no particular interest in minimizing the distance to the nearest stream, but they were resolute in seeing that this distance did not exceed a culturally defined threshold of approximately 700 m.

Water was a "privileged" attribute of settlement location. The reason for its influence was that with extensive cultivation offering high marginal returns to labor, fetching water from a distant source was the highest avoidable marginal labor cost. The normal human water requirements were boosted in this case by the heat in which the farmers worked and by the integral role of millet beer in the ecology.

As land availability tightened, the attraction value of a nearby source of domestic water dropped, with farmers becoming increasingly quick to pass up plots near the stream for being too small, too unproductive, or requiring too much intensification. Models of agricultural intensification (Boserup 1965) and of landscape attraction values (Chisholm 1979:94–103) may treat the costs of farm labor and nonagricultural travel independently, but the farmer weighs the two simultaneously and adjusts their weighting as land pressure changes.

Soil Type and Abandonment

The role of soil variability in locational decisions has changed in a different way through the course of the settlement pattern's development. The Kofyar may be aware of fine distinctions in soil properties, but, conforming to what is probably a common pattern (Wilshusen and Stone 1990), their settlement decisions on the early frontier were not very discriminating. Early bush farms were attracted more by availability of land than by edaphic qualities, and throughout the 1940s most migrant plots were located on thin and often quite rocky piedmont soils. When the bush farmers reached the deep sedimentary soils of the Benue Lowlands, their pioneering communities were located in soils of highly variable quality.

But whereas initial locational strategies were not sensitive to soil type, responses to land pressure were. As landscapes filled and yields were diminished by weeds, nutrient loss, and impeded drainage, agrarian settlement strategies diverged according to the marginal returns to intensification, with some areas being consistently abandoned and others developing stable populations of intensive farmers. The pioneers who had in the 1950s carefully chosen farm locations within 15 minutes of water and with little regard for soil distinctions watched the next generation compete for water-deprived areas where soil conditions were favorable.

Rules for What?

Throughout the study I have treated settlement decision making as inextricable from the process of agriculture, and I have gone to some lengths to describe an agroecological framework for understanding settlement. This framework is generally independent of cultural and historical context, and it builds on classic models of agriculture and settlement, such as those of Boserup and von Thünen, which were likewise acultural and ahistoric. The framework is based on the limited set of ecological relationships that crosscut cultural and historic contexts: extensive is more efficient than intensive cultivation, land pressure obviates extensive cultivation, increased labor access promotes farm-residence proximity, and so on.

This provides a framework for explaining the evolution of agrarian settlement systems. But it still does not yield a predictive model of agrarian settlement evolution. It is probably impossible, and certainly premature, to deterministically model the trajectory of agrarian settlement patterns in general because of the wide variation in cultural response to the ecological priorities and constraints. Land pressure's inhibition of swiddening may crosscut cultural and historic contexts, but responses to land pressure may differ sharply: one group might intensify agriculture, whereas another picks up and leaves and another takes steps to rid the area of some of its farmers. Field intensification's pull on the residence to the plot may be constant although responses to that pull vary through time, producing year-round dispersal in one period and nucleation with field houses in the next.

Yet I would not go so far as to agree with Grossman's (1971:23) conclusion, based on studies in West Africa, that "general 'laws' [of settlement] are meaningless outside the specific cultural and technological context." I think we can build settlement theory by first understanding the spatial ecology of agriculture and then seeking patterning in cultural responses to it. My study began with basic relationships between agriculture and settlement and then looked at how the development of one agrarian settlement pattern was played out in the context of those forces. The Kofyar pattern is not an optimal solution to the agroecology of the Namu Plains; it adapted to selective pressures but was not determined by them. An agrarian settlement pattern differing sharply in settlement size, occupation span, tenure arrangements, and land use could have developed in the same place, even by a population with the same agricultural technology as the Kofyar. To understand the context of the Kofyar pattern, it is essential to consider what sort of variation would have been possible on the Namu Plains, and ask why the Kofyar responded as they did.

Why *would* different cultures' agrarian settlement systems respond differently? In their own ways, anthropologists, sociologists, and geographers have all explored how adaptive ecology is shaped by cultural differences. Goldschmidt (1970) identifies the four elements of persons, rules, strategies, and goals. The first three elements are culture-free, but the fourth is not; we must determine what participants are "playing for," consistent with Weber's (1929) notion of substantive rationality (see Donham's 1990 examination of cultures maximizing situation-specific utility). If we are to think of different responses to ecological constraints as resulting from cultural "goals," we must ask what the possible cultural goals are and what explains the different goals.

Rural sociologists have proposed modalities in the cultural goals that shape agrarian settlement, with several studies converging on the concept of yeoman versus entrepreneurial farming cultures. Yeomen strove for community continuity, valuing long-term security over short-term profit. In the U.S. Midwest they were typified by German immigrant farmers, who were described as "plodding and industrious, methodically taking steps to establish sons on neighboring farms . . . concerned with continuity and traditions and unconcerned if economic progress came slowly" (Salamon 1985:325). In contrast, entrepreneurs were unsentimental cash croppers who acquired and abandoned land in pursuit of the bottom line. They were exemplified by British and Yankee (old American stock) farmers (Salamon 1985; Flora and Stitz 1985; Foster et al. 1987). These studies suggest that the yeoman/entrepreneur dichotomy results in patterned differences in key aspects of settlement pattern (such as occupation span, abandonment patterns, and methods of land transfer) as well as aspects of production that are integrally related to settlement (such as crop choices and stock ownership).

This dichotomy has the advantage of having been compared by several researchers to agrarian settlement data in separate areas, unlike, for instance, Williams's (1977:77) claims for different cognitive orientations affecting farm settlement of European pioneers in Argentina. However, there is no way to anticipate where else this particular dichotomy may be found. For instance, because it is based partly on participation in the market, is it restricted to cases where land is alienable and crops marketable? Furthermore, the reasons for the yeoman/entrepreneur divergence remain wholly unexplained, and it seems reminiscent of Kroeber's (1917) notion of a superorganic culture, detached from ecological and historic context.

The situation on the Namu Plains suggests another dichotomy that shapes cultural responses to general agroecological constraints such as those described in chapter 3. There are Tiv farmers on the same plains where the Kofyar farm, and there are intriguing contrasts in how the two settlement patterns have unfolded. Kofyar and Tiv may not represent anything like paradigmatic types. Nevertheless, the patterned differences in their responses to agroecological opportunities and constraints suggest configurations that recur cross-culturally, and I will show how the Kofyar-Tiv contrast is echoed in a very different frontier context.

Intensifiers and Extensifiers in Namu District

Tiv farmers were present in low numbers on the Namu Plains when the Kofyar began to settle there in the early 1950s. Some of the Kofyar ungwa were named for Tivs, including Duwe and Koprume (chapter 6). I encountered several cases of Kofyar establishing permanent farms on precisely the spots where Tiv had settled ephemerally. Indeed, the large compound where our research team lived in 1984–85 had been built and later abandoned by Tiv. I was intrigued by the permanency of Kofyar settlement in precisely the same locales where Tiv settlement had been ephemeral; differences between Tiv and Kofyar settlement are significant to my efforts at contributing to a general theory of agrarian settlement. Although there are no available studies of Tiv on the frontier, I can combine ethnographies of the Tiv homeland with limited data we have on the frontier to piece together a comparison.

Despite obvious differences such as language family (the Tiv tongue is Niger-Congo, the Kofyar is Chadic) and size of group (there are close to a million Tiv, and fewer than 75,000 Kofyar), the two groups have many parallels relevant to agriculture, including crops, agricultural technology, and level of market participation.[2] There were also initial similarities in settlement and land use. The Tiv built large compounds, some with several dozen huts housing multiple nuclear families; there were sometimes small compounds nearby. Aerial photos from the 1960s and 1970s show footpaths emanating radially, leading to cultivated plots as far as 500 m away. There were comparable macrocompounds among the Kofyar pioneers.

Localized population pressure developed for both groups. We do not yet know the extent to which the pockets of Tiv population were growing and the extent to which declining yields were necessitating longer and longer walks to fresh plots; a secondary factor may have been the decimation of local game. (It was the somewhat disdainful view of the Kofyar pioneers that the Tivs moved on when the bush rats had been hunted out.) At any rate, the Tiv response was often to abandon the locale and move on to a new spot on the nearly empty savanna.[3] Many of the abandoned localities, including Duwe and Koprume, were in the sand zone where the returns to intensification were relatively high. The Kofyar pattern, as shown in chapter 11, has been to sometimes abandon farms on soils that respond poorly to intensification and almost never to abandon farms on the sand zone.

This fundamental difference in response to land pressure has its roots in the Tiv homeland adaptation. In contrast to the Kofyar, the Tiv traditionally maintained a system of extensive cultivation by continually expanding onto new lands north and south of the Benue Lowlands. As Udo describes it (1970:142), the British sought to check Tiv expansion in the 1930s by building the "Munshi Wall" (Munshi being the Hausa name for the Tiv, used by colonial authorities), which Tivs simply climbed over.

Tiv expansion most often took the form of gradually encroaching on the land of other Tivs, with the outermost groups moving against non-

Tiv farmlands (Bohannan 1954a, 1954b). Tiv movement also took the form of disjunction, or emigrating from the area controlled by one's agnatic kin. Emigration was usually to other areas controlled by the Tiv, but by the 1940s the Pax Britannica allowed increasing movement into frontier areas. In location after location, land pressure brought not the heightened work of intensification, but movement.

Tiv movement in general is not only embedded in the agricultural system, it "is intimately associated with the social and political structure of the Tiv—is, in fact, a facet of it" (Bohannan 1954b:2). As described by Bohannan, and developed in an evolutionary context by Sahlins (1961), the hierarchical social organization called the segmentary lineage is singularly effective at mobilizing kin to contest the boundary disputes that inevitably result from constant encroachment on other farmers' lands. When a farm expands, it is against one's most distantly related neighbor, so that in the ensuing dispute, the antagonist will always have more closely related neighbors to call on. Dispute and the social relationships that structure disputes are ingrained in the settlement system.

Our understanding of this neatly articulated system of social organization, land tenure, and agriculture is based on ethnographic and ethnoarchaeological work in the Tiv homeland (see also Folorunso and Ogundele 1993). It is unknown if and how the system has worked among pioneering Tiv in the Namu Plains, but a system oriented toward "predatory expansion" (Sahlins 1961) would probably allow them to pursue a system of extensive agriculture and shifting settlement even in the face of considerable land pressure.

The Tiv provide an example of a different agrarian settlement pattern in the same environment as the Kofyar and show how the difference can be traced to different relationships among agriculture, settlement, and sociopolitical organization. Differences in settlement fission and agricultural expansion have been linked to social structure in other ethnoarchaeological studies as well (Kirch 1978:122). Might these relationships be organized into broader patterns? Let us pursue this possibility by examining a second instance of divergent agrarian settlement trajectories occurring in the same environment. The two comparisons offer a surprising set of parallels, despite sharply different circumstances, that will help to develop a more general perspective on cultural responses to agroecological constraints.

Intensifiers and Extensifiers in the American Backwoods

The second case, set in the 17th-century eastern United States, is the subject of a historical ecological study by Jordan and Kaups (1989). These geographers describe two very different agrarian settlement systems in the Delaware Valley in the eastern United States. The differences in these adaptations trace to their roots in Europe, where there existed what might be called an ecological core and periphery. The fertile Germanic core of Europe, with its stable (and partly intensive) three-field system of production, had a prosperous, conservative agrarian population. Fringing the core were British hilly areas, the subarctic north, and the infertile Eastern European plain. Population here was sparse, and land productivity low. This area gave us Scotch-Irish, Welsh, alpine Swiss, and Finnish settlers. Of particular import was an adaptive system originating in eastern Finland, with its short growing season, thin morainic soils, and pine-spruce forests. In the *huuhta* farming system, cultivation was very extensive, with rye being grown in the ashes on swidden plots that were abandoned after a single year. The Finns kept open-range stock and also hunted and gathered. This system relied on an estimated 2500 acres per family. They lived in one-room cabins or small multistructure farmsteads, which were usually abandoned every few years.

This highly extensive agrarian settlement system was banished to America when it started to encroach on the ecologic core. The swiddeners had colonized most of interior Finland and moved into Sweden and Norway in the 1600s. The valleys here supported stable, intensive-farming villages, and the Finns colonized the interfluves very successfully, sticking to the high ground, until "the Germanic Valley folk awoke to find Finns perched in the heights above them, and the new ethnic map of south-central Scandinavia had topographic lines as borders" (Jordan and Kaups 1989:51). Laws were enacted to constrain the wanton destruction of forest and game, and many swiddeners, perceived as landless vagabonds, were rounded up for transport to the rich deciduous eastern woodlands of America ("a bit like tossing Br'er Rabbit in the briar patch," as Jordan and Kaups [1989:58] wryly note). In the 1650s a Finnish population was in place along the Delaware River, where the Finns thrived—hunting, farming Indians' land, being imprisoned for wizardry, brewing beer and vodka, and farming very extensively.

The agrarian settlement system was preadapted to the woodlands, and it flourished in the backwoods frontiers where European settlement had yet to reach. Farmsteads with a log cabin and sometimes an outbuilding or two were established near old Indian fields that provided forage and attracted deer. The main agricultural job was tree chopping, a task to which the Finns brought not only technologically superior axes but skill at their use, and an ethic that chopping was preferable to sod-busting. They also chopped trees for wages, and it was Finnish axmen the Dutch hired to clear Harlem in 1661.

Influenced by Indian agricultural tactics, they intercropped corn with squash, pumpkins, or watermelons in fields where the tree stumps were left standing. Vegetables and tobacco were grown in kitchen gardens. The only tree crop was peach, which bore fruit within three years, required less care than apples or pears, and produced brandy. They relied significantly on hunting and gathering.

An essential component of this adaptation was mobility. Driven by what Jordan and Kaups call a "cultural resistance to intensification of land use," the backwoodsmen dispersed rapidly along watercourses. "The backwoods folk had to choose between intensifying land use and preserving their traditional lifestyle through migration. Some contemporary observers, core Germanic types all, could not understand the motives of those who moved, believing instead that migration 'prevented their enjoying the fruit of their labours.' . . . But in cultural ecological terms, to remain would have meant to adopt a new adaptive strategy, whereas the only long term payoff in adaptation is to continue to function in the traditional manner for as long as possible" (Jordan and Kaups 1989:78).

The clearings won from the forest by the pioneers served them at best for three to five years. Crops thrived upon the burst of fertility derived from the ash, and regardless of how natively fertile the soil was, yields fell off rapidly after several harvests. In addition, weeds quickly became a problem that could be controlled only by a large investment of labor. The backwoods pioneers had no ability at or interest in restoring the productivity of the soil through manuring, and they refused to submit to a season of hard hoeing. Let their Germanic successors accomplish such intensifications of land use; the Midland pioneers preferred to make another clearing in a different place (1989:100). The Finns, Scotch-Irish, and others from the European periphery had in effect "skimmed the cream" from the woodlands; the price had been a high degree of mobility, which bothered them not a whit. In fact, for short moves, cabins could simply be dismantled and reassembled in the middle of a new field.

The second wave of settlers was obliged to farm more intensively, to which they too were preadapted because they came from the ecological core of Europe. These more permanent settlements varied from dispersed farmsteads to small villages, but they shared a key set of characteristics that contrasted with those of the Finns who had come before. Most important, when faced with the choice of intensifying production or moving on, they intensified. This meant planting on bush-fallowed land, which meant plowing, which meant the arduous work of stump clearing that the Finns had disdained.

Culture and Ecology

Jordan and Kaups present the Fenno-Indian swidden system as an example of the "pioneer farmer" stage in Turner's history of the American frontier and as an example of a singularly successful strategy of forest colonization (1989:7). They are critical of what they call normative models, such as Green's (1979) model of temperate forest colonization, which levels differences among cultures and winds up with the rather awkward prediction of parallel adaptations in all temperate woodlands. Jordan and Kaups call their work "particularistic cultural ecology," a framework they believe anthropologists have discarded too hastily. In this framework, separate adaptive pathways result from the interplay between the unique history of a culture and its physical environment. Jordan and Kaups are explicitly interested in the particulars of this case, and they cite as a goal of cultural geography the gaining of "a better grasp of general process so as to understand an immensely complicated history" (1989:19).

Because many archaeologists and other social scientists envision the opposite goal—a better grasp of complicated histories so as to understand general process—it is worthwhile to try to fit these cases into ecological principles. It is true that the first European settlers in the American Midlands and the Tiv in the Namu savanna thrived precisely because they were preadapted to the system of extensive farming and rapid abandonment that frontiers select for. Margolis (1977) shows that this favored frontier adaptation conforms to the strategy of "fugitive species" in newly opened habitats. The extensify/abandon strategy thrived on both early frontiers, whereas the intensifiers dominated the second stage, with their enduring settlements. The concept of cultures adapting to separate ecological niches (Barth 1956) is especially apt when the niches open serially, as in ecological succession (cf. Gall and

Saxe 1977); in both of these cases, the intensifiers' niche was improved by the field clearing done by the extensifiers before them. It is equally consistent with economic models of land-use competition (Hudson 1977).[4]

But there is more to it than this. The Tiv and Finnish agrarian settlement systems may have been especially adapted to frontier conditions, but some of the Germanic and Kofyar farmers were there in the same conditions, at approximately the same time. I would say that both the intensifying and the extensifying systems make ecological sense, but only if we see the agrarian settlement system as embedded in social organization and labor mobilization.

In cultural ecology, causality tends to run from the effective environment to food acquisition systems to cultural domains such as social organization and settlement (Steward 1955; Flannery 1972a). Food acquisition tactics are selected according to their efficiency or marginal utility, intrinsic properties that are not culture specific. Thus, intensive agriculture is seen as less efficient than extensive agriculture, regardless of the farmers' culture.

This does not mean, as Bennett pointed out (1967:21), that cultural forms do not influence adaptation. All strategies of food procurement and production require technology, including what we might call social technology—conventions for mobilizing human resources. Social technology that facilitates a food production strategy lowers the cost of that strategy. Earle writes that "because social relationships play an important role in procurement strategies, the possible forms for the organization of labor in exploitative tasks are set by a group's structure . . . as well as by the technical requirements of the work. To the degree that social structure affects the possible organizational forms, it affects the costs of exploiting a resource and, thus, its importance in the subsistence economy" (Earle 1980:4; see Ensminger 1992:169). Thus, efficiency of production strategies can vary culturally, and even a purely ecological analysis of basic strategies such as agricultural intensification versus abandonment needs to consider social (or informational or even ideational) technology that affects costs and benefits.

The intensive agriculture that the Kofyar make look easy is remarkably difficult, not just because of the gross labor requirements but because of the confounding scheduling problems that arise in a system where the pivotal fuel of intensification is human labor. The Kofyars' "culture," or at least their experience over the past century, effectively lowers the cost of intensification via an intricate array of conventions for mobilizing labor (Stone et al. 1990), their reliance on the household as the basic productive unit, and even some of their fundamental values.

At the same time, abandonment and residential mobility were not a part of the traditional Kofyar system. In the Jos Plateau homeland, a man usually changed residences only when leaving his natal household; if he inherited the family farm, he might never move. Not surprisingly, social technology provided little support for residential mobility. The Kofyar household was too small to act as a pioneering unit on the Benue frontier, and would-be pioneers had to persuade other households to join them in forming extemporaneous migrating groups. Land disputes had traditionally been handled by village elders or clan members, procedures of no use when disputants were non-Kofyar. Intensification, which we usually see as the expensive way out, was cheaper for the Kofyar because of their social technology.

For the Tiv, social technology lowered the cost of abandonment by facilitating migration and expansion into new land. The segmentary lineage provided a ready principle for organizing pioneering units that settled in the Namu area early on. The lineage system was adept at mobilizing support for land disputes with non-Tiv and Tiv alike.

The contrast finds parallels on the American backwoods frontier and other multiethnic frontiers as well. It shows how land pressure may demand the end of extensive farming but not dictate whether or not farms are abandoned. In many cases it is impossible to model farmstead abandonment without knowledge of (or assumptions about) other aspects of culture, which may include social organization and ideology. In order to understand abandonment, then, we must put it not only in the context of the agrarian settlement system, but in the context of fundamental questions of causality that have traditionally divided anthropologists.

The interpretive/idealist tradition in anthropology sees culture as the process of making collective sense of the world, especially by manipulation of symbols; the materialist/ecologic tradition focuses on physical interaction with the world, especially the acquisition of basic provisions. At least a few archaeologists have sought a common ground; Flannery, for instance, has long advocated the inclusion of cultural management of information in ecological models (Flannery 1972b:400). But in exploring the mediation of idealist and materialist causality, Flannery has tended to focus on isolated fragments of ideology, such as the Zapotec concept of good farmland (1986:516) or the Andean ideas about propitiation of ancestors (Flannery et al. 1989). Instead, developments in the

Namu savanna and the Delaware Valley reflect organic linkages among culturally instituted agricultural tactics, residential mobility, social organization, and ideology.

These case studies suggest separate intensifying and extensifying modes of agricultural adaptation. My examples of frontier intensifiers (the Germanic types and the Kofyar) come from intensive-farming traditions, and the extensifiers (the Finns and the Tiv) come from extensive-farming backgrounds—but my point should not be mistaken for cultural reductionism. It is rather a recognition that on these frontiers, there was no single factor that dominated the unfolding of settlement patterns; there was no good analog for von Thünen's transport cost, Chisholm's time cost, or Christaller's efficiency of distribution. The common denominator on these frontiers is that as population grows or resources degrade, the farmer looks ahead to an intensification slope on declining returns to marginal inputs. The point at which the farmer abandons the farm (and the local intensification slope with it)—or takes other action to reduce the pressure to raise production—is influenced by various factors extrinsic to the slope. There are technological factors: the Finns' incessant chopping was underwritten by their superior axes and their skill at using them. There are social organizational factors: the Tivs' segmentary lineage was ideally suited to squabbling over land they encroached upon, whereas the Kofyars' labor organization was adapted to the manpower demands of intensification. These are basic material conditions that affect the costs and benefits of abandonment as opposed to intensification. There are also cultural values, strangely similar in these two very different frontiers: the Finns' disdain for the Germanic system, with its stumpless fields and tidy villages, finds an echo in the pioneering Tiv climbing over the wall with hoes slung over their backs, never stopping to question the cultural dictum that each farmer had the right to the land he needed.

As I have pointed out elsewhere (Stone 1993a:79), the idea of two distinct modes of adaptation—such as intensifiers and extensifiers—is clearly a simplification. But models are by their nature simplifications, and perhaps this one helps us to see a little better where we need to go from here.

Appendix

Methods of Studying Agrarian Settlement

This section outlines the primary sources of data on which the following discussions of Kofyar frontier agriculture and settlement are based. These data were collected between January 1984 and February 1985 by M. Priscilla Stone and myself, with Robert McC. Netting participating between January and July 1984. I have also drawn throughout on data collected by Netting in his prior fieldwork. We used a wide variety of survey instruments, and I will limit myself here to those that played the greatest role in the analyses in this book.

Kofyar Settlement Survey

In 1963, Nigeria was aerially surveyed by Hunting Surveys, producing a series of 1:40,000 photographs of variable, but generally good, quality. The photographs of the Kofyar frontier area show land use fairly clearly but even with the use of a stereoscope with an 8X eyepiece permit only some individual compounds to be resolved. Topographic maps have been compiled from these aerial photographs and were used extensively in this study.

In the early 1970s the Kofyar area was again photographed, as part of the Land Resources of Central Nigeria study carried out by the Land Resources Division of the British Ministry of Overseas Development. The Merniang portion of the Kofyar homeland was photographed in 1971, and the Doemak, Kwalla, Namu, and Kwande areas in 1972. These images are at the scale of 1:40,000 and are of high quality.

During the rest of the 1970s, Nigerian Federal Surveys began to photograph the entire country at the scale of 1:25,000. The Kofyar homeland and the Namu-Kwande areas were photographed around 1978. This imagery, of large scale and for the most part very high quality, has been used as the basis for the spatial analysis of Kofyar settlements.

The 1978 aerial images clearly resolve individual compounds and with a magnifying stereoscope or computer enhancement even resolve individual huts. These photographs were used in the field to construct a Cartesian database of Kofyar settlement. Transparent overlays were affixed to photographs within the study area, and features such as compounds (both inhabited and abandoned) and other features were identified by ground surveys. Surveys were conducted after an ungwa had been censused. Compounds were usually identified with the help of an ungwa resident who walked with my field assistant and me. When no local helpers were available, we would visit each compound in turn and get enough information about the household to determine which census pertained to the compound.

As each compound was identified, it was circled on the overlay and a code was written next to it. The household census numbers for the compounds—along with information on changes in settlement, land use, and landscape features that postdated the photos—were spoken into a tape recorder as we moved through the ungwa, and these were later transcribed.

Compounds erected after 1978 were recorded in one of two ways. When the surrounding landscape contained distinctive landmarks so that the location of the compound could be determined accurately on the photograph, the compound was recorded with a dot on the overlay. When this was not possible, I used a Brunton compass and distance wheel to traverse from a clearly visible point on the aerial photograph to the compound. Coordinates for compounds mapped by traverse were computed in the field with a Hewlett-Packard 41CV programmable calculator, using modified versions of programs in Hewlett-Packard's Surveying-I software module. The settlement overlays were later combined into a master settlement map, and settlement loci were digitized on a graphics tablet. These settlement records became a part of a relational database, linked to the census and labor data described below.

A second spatial database was compiled for analyzing relationships between settlement and landscape features. An overlay was made from the 1977 aerial photographs, on which each stream was digitized as a series of points. Care was taken to ensure that a stream point was digitized wherever the stream came closest to a settlement, to allow accurate computation of distances to nearest water. Later, the digitized boundaries of soil zones were added.

Finally, to determine changes in settlement patterns recorded on aerial photographs from different points in time, I directly compared images shot in 1963 and in 1978 with the use of a Zoom Transfer Scope. This allowed me to identify compound abandonments and to recheck household settlement histories.

Household Census

A household census form was used as the principal research instrument in the 1984–85 fieldwork. The unit of data collection in the census was what we may loosely call the household. However, in contrast to the traditional settlement in the hills—where the household was a discrete social, economic, and co-residential unit—frontier farming households may occupy two compounds on the same farm, or two or more separate farms; there are also cases of two economically separate households residing in a joint compound. The Kofyar define the household as those individuals, whether co-resident or not, under the authority of a single household head. The emic household may be spread over two or even three farms; it may include a home farm that the frontier members almost never visit; it may even include a frontier farm worked by an economically independent adult son and his family, who still defer to his father as wupinlu.

We used the emic household as the data collection unit because of its logistical advantages but collected information so that spatially and economically discrete units within the household could be separated in analysis. For instance, we recorded where each individual spent what proportion of his or her time, so that the personnel for specific farms could be isolated within the household.

The census form comprised two parts, the first concerned mainly with individual members of the household and the second mainly with agriculture and settlement. For each individual we recorded basic demographic data as well as information on their activities: any economic pursuits other than farming, and which farm(s) the individual worked on. Marital histories were collected for each woman. The wupinlu was asked for a settlement history and an anecdotal account of key aspects of the developmental cycle of his household (described below). Agricultural production was recorded both for the household and for those members that farmed for their own accounts. A sketch map was made of the compound at the farm where the census was conducted, and details of architecture, use, and age were recorded for each structure (when possible, we later collected the same information on the compounds on the household's other farm). Also recorded was domestic animal ownership, sponsorship of group labor, and the hiring of wage workers during the previous year. The census generally took between 30 and 90 minutes to conduct.

Censuses were conducted by each of the investigators and by our bilingual Kofyar assistants. Each census collected by an assistant was carefully checked and discussed with one of the investigators. When census data were deemed inadequate, we were usually able to revisit the household.

Labor Study

To measure labor inputs into farming and to elucidate the spatial organization of agriculture, we arranged to have records kept on the daily activities of each member of a sample of 15 households. Seven were in Ungwa Kofyar, 4 in Kwallala, and 4 in Dadin Kowa. Locations of the Ungwa Kofyar and Kwallala labor study households are shown in figure 9.1. Our local enumerators recorded crop, task, estimated hours, and type of labor mobilization for 26 males and 36 females over age 14 for a total of 50 weeks. As we have stated elsewhere (Stone et al. 1990), "Since the enumerators had to be knowledgeable, resident observers, reporting on their own households and in most cases also on a neighboring household, it was impossible to use a probabilistic sample of households. However, the sample households are representative in size, averaging 4.13 adult workers (ages 15–65) as compared to [a] mean of 4.20 for the frontier households in our larger census (n=865). The sample households were located so as to sample the spectrum of population density and agricultural intensity." Enumerators made daily entries for each adult, relying on individual reports for tasks not personally observed. Time spent on each activity was rounded to the nearest hour, and activities less than approximately 45 minutes in duration were omitted. This procedure produced an underreporting of domestic tasks, but almost all agricultural work is at least 45 minutes in duration. The locations of almost all off-farm agricultural labor bouts were identified in my settlement survey, allowing the reconstruction of movement and travel patterns in detail.

Computer Analysis

Data were managed by software written by myself (in Turbo Pascal) specifically for the Kofyar project. Data entry and management were handled primarily by a relational database system called AddRec, which contained separate arrays on households, farms, individuals, farm production, settlement histories, marriage histories, group-labor parties, and residential compounds. I also wrote an analytic mapping program (SetMap) that displayed selected groups of settlements, using color and symbols to show other variables such as village and region of origin, sargwat affiliation, residence time, and aspects of settlement history. AddRec could extract flat files (rectangular arrays) that were then imported into statistical packages, spreadsheets, surface analysis programs, and my own programs for analysis. The great majority of the data analysis was done in spreadsheets.

Notes

Chapter 1. Introduction

1. Nomothetic explanation attempts to subsume phenomena under laws or lawlike statements; see Binford 1968a; Watson et al. 1971; Salmon and Salmon 1979.

2. Flannery actually considers the rules to be only a guess (1976a:180), although others have treated them as being more conclusive (e.g., Damp 1984).

3. For instance, Bylund 1960; Morrill 1962, 1963a; Hudson 1969; Hamond 1981.

4. As in, for instance, Keegan and Machlachlan 1989:614.

5. See, for instance, Shyrock 1939; H. Johnson 1941, 1945, 1951; Conant 1962; Lemon 1966; Eidt 1971; Jordan 1976; Friesen 1977; McQuillan 1978; Jordan and Kaups 1989.

6. For instance, Bylund 1960; Morrill 1963a; Sanders 1967; Mortimore 1968; Cordell 1975; Jordan 1976; Madsen 1982; Voorhies 1982; Haberland 1983.

Chapter 2. Causality in Agrarian Settlement Systems

1. Territory is sometimes defined as "the area habitually exploited from a single site" (Higgs and Vita-Finzi 1972:30) and catchment as the entire area from which a site's resources are derived (Jarman et al. 1982). The former concept is more relevant to agricultural location theory and empirical studies of agricultural movement.

2. In fact, it is unclear that the partial fit between the predictions and the test case in Iowa was due to the causes Hudson proposed. More recent studies of the Great Plains have shown different historical patterns of changing farm sizes to be due in part to differences in local ecology (Baltensberger 1993), a factor held constant by Hudson.

3. U.S. and Canadian laws in the 1860s encouraged settlement of the West by providing cheap 160- or 320-acre farms; this was based on what could be worked by one man with a team of horses in the East with diverse crops (Bennett 1969). But this farm size was too small for successful grain ranching, and many farmers could not build up a cushion for bad years (Netting 1993:152–153). Droughts and recessions drove some out, and the remaining farms expanded. In Saskatchewan, mean farm size rose from 432 to 686 acres between 1940 and 1960.

Chapter 3. Agrarian Production and Settlement

1. This is a common usage of the term *efficiency* in agricultural ecology (Brookfield 1972), not to be confused with the economists' concept of market efficiency.

2. Turner and Doolittle (1978) show how capital and labor inputs can be incorporated into a single model, paradoxically classifying some systems with low inputs of human labor as intensive.

3. Turner et al. (1977), in a study of tropical subsistence cultivators, found that population density accounted for 58% of the variation in agricultural intensity (if one problematic case is removed, this climbs to 67%). Evidence from sub-Saharan Africa has provided much support for the Boserup model—protestations that it does not fit Africa notwithstanding (Guyer n.d.; Kalipeni 1994). W. B. Morgan's (1955b) study correlated fallow shortening and population in Nigeria, and Gleave and White (1969) found the same relationship in their appraisal of West African agriculture. Case studies on the Nigerian Kofyar (Netting 1965, 1969) and Tiv (Vermeer 1970) and the Tanzanian Matengo (Basehart 1973) fit the model well, as did a survey of cases from sub-Saharan Africa by Pingali et al. (1987). A collection of detailed African studies also sustained the pattern (Turner et al. 1993; Kates et al. 1993), as did controlled comparisons in various regions of Nigeria (Lagemann 1977).

Also supporting the model are cases where population decline prompts a reduction in agricultural intensity. The Kofyar, for example, left a crowded homeland with intensive agriculture for a sparse frontier where shifting methods were adopted (Netting 1965, 1968); in the remaining homeland villages, where population was sharply reduced, cultivation was also extensified (Stone 1996).

4. My point here concerns the *shape* of the intensification slope—i.e., relative rather than absolute values—and therefore the chart is not scaled. I will not deal with the knotty issues in measuring input-output efficiency. However, the input measurement would have to include indirect costs such as travel time to fields, the cost of agricultural infrastructure development and maintenance (of terraces, dikes, corrals, etc.), and part of the cost of activities that enhance production (tending manure producers, brewing beer for festive labor, etc.). The inputs, as Turner and Doolittle (1978) have pointed out, need not be only the hu-

man labor stressed by Boserup but may be capital as well. Increasingly capitalized intensification tends to involve increasingly indirect costs which are difficult (although not impossible) to factor into the intensification slope model. For this reason I am using the model only to depict the dynamics of agriculture without substantial capitalization

5. The end of the intensification slope indicates a point beyond which production concentration cannot be further increased with available technology. This means that the production regime of agriculture does in theory have its own Malthusian endpoint. But Robinson and Schutjer (1984) and Lee (1986) show how the entire spectrum of agricultural intensity can be seen as a node in a larger intensification sequence that may include other nodes such as industry.

6. Locally implemented intensification slopes provide a framework for understanding the effects of environment on settlement, for which models have fared poorly over the years. Demangeon (1927), for instance, proposed settlement rules predicting dispersion in mountainous areas and aggregation on plains; Semple (1932) explained settlement patterns in arid areas as a simple function of water resources. Blok (1969:125) later tried to include more variables, including the effect of water supplies on agglomeration, but this scarcely improved the model. What was needed was a recognition of the dynamic role of agriculture in mediating the relationship between environment and settlement.

7. She "appears to exclude the possibility of farmers under population pressure expanding into uncultivated areas. . . . [I]n the model, the supply of land which can be cultivated is fixed" (Grigg 1979:69–70). This allows her to isolate a causal relationship between population density and agricultural production, but at the expense of understanding the relationship between intensification and the spatial arrangement and mobility of human population. Nevertheless, Boserup was not oblivious to mobility, as her description of shifting settlement makes clear (Boserup 1965:70–71, 1970:101).

8. My concern here is with the effects of agriculture on settlement, but extensive agriculturalists are characteristically hunters, and game availability may also be a consideration in settlement shifting. This has been best documented in the Amazon (Gross 1975; Hames and Vickers 1982:363), although the Amazonian data also underscore the importance of agricultural factors such as soil depletion (Gross 1983) and the distance to well-rested soils (Vickers 1988) in village movements.

9. The pull toward the plot varies not only with cultivation intensity, but also with the crop. Root crops, for instance, because of their weight, have a higher transport cost than grains.

10. The issue here is with the relationship between crops and settlement; for a parallel relationship, see Rappaport 1968:70.

11. A general association between nucleation and scattered holdings, as well as between dispersal and compact holdings, had long been recognized, but there

was no unifying theoretical context for this until Chisholm's emphasis on distance separating the settlement from the field (Brookfield 1968:426). That agricultural strategies have vital ramifications for settlement organization has been widely missed, as in Blok's (1969:132) attempt to predict dispersion and agglomeration: "[I]n pacified areas agglomeration will persist, if labour on the land is frowned upon and if resources are strongly limited."

12. There is disagreement over this causal relationship in the settlement literature. Richardson, for instance, in a study Chisholm uses in the latest edition of *Rural Settlement and Land Use* (1979), compares rice farmers who live away from plots they single-crop and farmers who live on or near fields they double-crop. Richardson concludes that "this locational proximity to the ricelands encourages high labor inputs and high yields" (1974:241–242); however, the causal arrow generally points the other way.

13. In fact, in the "ethnographic present," systems of shifting settlement with shifting cultivation are exclusively an adaptation to tropical ecology. All 15 cases of shifting settlement in the *Ethnographic Atlas* are located in the tropics, i.e., at latitudes between 23°N and 23°S (Ruthenberg 1980:1).

14. The patterns I am describing should have some applicability to those tropical groups that subsist on both extensive farming and hunting-gathering. Vickers's (1988) longitudinal study of the Ecuadorian Siona-Secoya shows that despite a substantial reliance on hunting, game depletion plays a smaller role in village abandonment than do horticultural transportation costs and deterioration of village structures.

15. The Esch was an intensively farmed, annually fertilized infield area. Characteristic of northwestern Germany, it was divided into strips for use by individual farmsteads that were usually located around it in what was called a *Drubbel* settlement (Mayhew 1973:18,219; Dickinson 1949:244; see Eidt 1984:12–21 for a review of German settlement research).

16. There are also problems with population measurements for the Dugum Dani; Heider (1970:60) states that "the population figure is only a guess and may be off by 50 percent." The Gururumba are tied for the fourth highest population density in the sample and are classified as living in villages. Brown and Podolefsky (1976:237) use a figure of 44/km^2 (113/mi^2), whereas the ethnographer describes Gururumba as "1121 people living in six large villages in the northern half of a large bay . . . a territorially distinct unit of approximately thirty square miles" (Newman 1965:30). The latter figure works out to 14.4/km^2 (37/mi^2), the third *lowest* population density in the sample of 17 societies. For a more recent analysis see Shankman 1991.

17. The quality of work by collaborative parties is variable. The festivity that helps attract workers may also promote carelessness (Erasmus 1956:456–458; Saul 1983), although work quality is quite high in Kofyar labor groups. Communal work groups may also offer greater dependability and lower transaction costs than do hired workers (Ensminger 1992).

Chapter 4. The Kofyar Homeland

1. There is no convincing evidence for Unomah's (1982) claim that Kofyar (Kwalla) settlements were near the Benue making salt before 1700.

2. To the south and east, respectively, are the hamlets of the Njak and Doka, who share language and origin myth but differ in settlement and some aspects of food production.

3. It is likely that some inhabitants of Kwalla were from a different cultural group, and early British administrators believed them to have migrated from the northern Jos Plateau in the 1830s (Temple 1919:253).

4. Netting points out (1968:135) that additional workers can be put to work on outfield plots, but outfield plots are scarce on the plains where the largest households are. Short on land and long on workers, these households were prime candidates for migrant farming; plains households led the diaspora (Stone et al. 1984).

5. *Mar* means farm, both the verb and the noun; *muos* means beer. *Mar muos* can be translated either as "beer farming" or as "farm (or field) of beer."

6. Netting (1968:99) provides average farm sizes for three hill communities and one plains community, but there are marked differences among hill communities and a paucity of measurements on the plains. I also suspect that the plains community of Dunglong, Kwa, where he made his measurements, was somewhat atypical in having nearby uninhabited lands that could be brought into cultivation when needed (see Stone 1994).

7. This is based on a sample of 477 households for which Netting collected information on huts. Data from 1961 provide a fairly good picture of the settlement pattern just before the movement to the frontier; approximately a quarter of the Kofyar households were involved in frontier farming, but virtually all kept the homeland farm as their primary residence.

8. Netting (1968) has treated plains settlements as conforming to the same village/toenglu structure as in the hills (Netting 1968). But many of the plains "toenglu" settlement groups are larger than entire hill villages. These differences in size are gross enough that we should expect differences in how toenglu and villages function in the hills and plains. These differences were clear in the development of frontier settlement organization. For instance, there were cases of new frontier villages being founded by farmers all from the same plains "toenglu," which could not have been done by the 20 or so adults from a typical hill toenglu.

9. *Plains Merniang* refers to the sargwat comprising the Kwa-Kwang-Fogol area, and *Merniang* refers to the "tribe" comprising the Plains Merniang and Gankogom alliances. Similarly, *Plains Doemak* refers to the sargwat at the escarpment base, and *Doemak* refers to the "tribe" comprising the Plains Doemak and Ganguk alliances.

10. *Gankogom* means "the *harmattan* side," referring to the dust-laden winds

that migrate annually from the Sahara, enveloping the area during the dry season. *Ganguk* means "the lee side."

11. Opinions vary on the propriety of the term *tribe* (Gandonu 1978; du Toit 1978; Sithole 1985), but the alternatives seem to pose more problems than they solve (e.g., Bromley 1978). I use the term here simply to label a group within which there is social propinquity relative to the surrounding populations. Patterns of social propinquity may be quite different from political organization and self-concept (Cornell 1988).

Chapter 5. Frontiering

1. The soils in the Benue Piedmont have been identified as ferric acrisols with dystric nitosols, and the soils in the northern portion of the Benue Valley have been identified as dystric nitosols with ferric luvisols (LRDC 1981:map 6). The rocky piedmont corresponds to the LRDC's land system 587, and the smooth piedmont corresponds to land system 588. The Namu Sand Plains correspond to the LRDC's land systems 567 and 582. The Jangwa Clay Plains correspond to land system 572.

2. The Goemai were called the Ankwe by the British, who established a divisional headquarters in the major Goemai town of Shendam. Although Shendam, Kurgwi, and Kwande belong to the modern Goemai "tribe" by any reckoning, the tribal model of ethnicity breaks down farther to the west. The small nucleated settlements of Shindai, Njak, and Namu share the Kofyar origin myth but resemble the Goemai in settlement pattern and agriculture. Namu had ritual relations with the Kofyar.

3. Parts of the Namu wall were still visible in 1985. In 1961, Netting was told that the wall had been built by the chief's grandfather for protection from Doemak and Kwande. He was also told of slave raiding by the Jukun.

4. Early in the colonial period, Lord Lugard wrote that "my sympathies are largely with these ignorant pagans, whose attacks on traders are often prompted by a natural retaliation for the enslavement of their relatives" (Protectorate of Northern Nigeria 1901:264). However, as the years wore on, the British came to regard some groups with considerably more sympathy then others. In 1945 Findlay described the Doemak (Dimmuk) as "a cheerful people, very fond of their beer, and very different from the dour, hard-headed Montols, another tribe in Shendam Division with a history of violence behind them" (p. 139).

5. A food raid by Koeper (the northernmost Kofyar hill village) on the plains village of Leet occurred in 1918 (PRO 1918:File 44654).

6. The mining of tin deposits on the plateau, which had preceded the colonial period, was expanded rapidly by the British beginning in 1902. In 1914 a railway was finished that connected the tin mining area to Zaria (W.T.W. Morgan 1983:115).

7. British officials frequently equated administrative control with direct taxation (Netting 1987:374). In 1918, Governor-General Lugard celebrated the peaceful collection of increased taxes from the Tiv as "a wonderful example of the genius of our race for controlling subject races" (Perham 1968:524).

Chapter 6. Pioneer Agrarian Settlement

1. The map is based on Hunting Survey's 1:40,000 air photos, corroborated in some cases with ground survey. The photographs were examined with a magnifying mirror stereoscope and by computer enhancement of high-resolution scans. Still, some of the locations of residential compounds are tentative because of the small scale and lack of sharpness.

2. In depopulated Kofyar hill villages, remaining farmers often move into compounds in a limited area for the same reason.

3. Doedel's part of the settlement has long since been razed, but the site is still marked by an enormous borrow pit that supplied the mud for construction. Nearby stands the compound of the present chief of Lardang (the nephew of Baam) and, across the road, the large compound of Garba Wubang.

4. Kofyar compounds do not suffer from the problems of vermin infestation and structural decay that lead to abandonments in many areas (e.g., Vickers 1988).

5. In depopulated hill communities in the homeland, farmers now must contend with animal encroachments. Visiting the nearly abandoned village of Koepal, I heard a young girl loudly singing from the top of an oil-palm tree, acting as a verbal scarecrow to ward off monkeys. In some areas people are giving up the cultivation of millet, which is eaten by monkeys. The factors encouraging Kofyar macrocompounds were quite comparable to those of the Thai frontier, where settlers joined 4–7 longhouses to collaborate in field clearing and protection from jungle beasts (Moerman 1968). Yet depending on the nature of the threat posed, animals may be a force for dispersal. In Sarawak, Iban built subsidiary houses—near their plots and away from the longhouse—to protect the crops against animal predation (Freeman 1955:33).

6. I believe the Tiv settlement near the southeast corner of figures 6.1 and 10.1 is that of Duwe.

Chapter 7. Land Pressure and Intensification

1. Allocation of time to different pursuits and locations is measured on a six-step scale, ranging from full time to occasionally.

2. This figure is based on two sources. First is my analysis of 1978 1:25,000 aerial photographs, indicating that 65.1% of the core area was in crops at that time. However, land-use intensity must have increased somewhat between

1979 and 1984. Second is my recording of land use along a 10 km stretch through Ungwa Long, Wunze, and Ungwa Kofyar in mid June 1984, showing 80.3% of the land to be in crops. The roadside land use will be higher than that of the area as a whole, because compounds are attracted to roads and land use tends to be most intensive near the compound. The information from the two sources is therefore consistent, and 70% is a good estimate of the amount of land in crops in 1984. Approximately 3% of the land is made up of roads, streams, and compounds, and I would guess the remaining uncultivated land to be divided evenly between fallow and uncultivated savanna.

3. This is based on a roadside land-use survey in September, showing 44.3% of the land to be in crops. This survey avoided some of the problem of infield overrepresentation, but it was conducted after some of the peanut and bambara nuts had been harvested (see Stone et al. 1990:17). Because these are especially important crops in Mangkogom, the figure of 44.3% should be raised to approximately 50%.

4. Our neighbors routinely requested samples of vegetable seeds that M. Priscilla Stone and I had bought back from Jos—despite our lackluster vegetable garden, which failed to produce a single meal.

5. This figure differs slightly from the 1599 (1661 including brewing for work parties) reported by Stone et al. 1990, after the following adjustments. First, I rewrote the program for extracting data from the database so that it is more sensitive to differences in tasks and crops. Second, I excluded crop processing work from this analysis; in my theoretical approach agriculture affects settlement through the cost of moving to one's field or to other farms, and processing is usually done within the household by household members. However, these figures do include brewing for work parties. Third, during the first two weeks of June, when my colleagues and I collected data, the coding and checking procedures were standardized, and the values were anomalous (see Stone et al. 1990: fig. 2). I have replaced these values by assuming a straight line between the values recorded in late May (these were actually collected in the 1985 season) and mid June. Fourth, as in the 1990 study (Stone et al. 1990), I have assumed that labor inputs during the two dry-season weeks when the enumerators for the study were on break were 80% of average.

6. There are several minor formats for mobilizing labor that I have lumped with wuk, including *zumunta* (church-group work details) and *mar lua* ("meat farming," in which a group of young men work for a meal with a meat entree). These miscellaneous formats constitute less than 2% of the total labor budget. Wage workers were hired by 36.2% of frontier households in 1983. The average number of paid person-days was 12.0 (s.d.=30.4, n=851), which is approximately 1% of the total 1984 labor input. Data on wage work were collected as part of an agricultural census, and the precise timing of the wage work is not available. Wage labor is not included in this analysis.

7. However, I would expect labor-pooling strategies to change through time. For instance, in 1992 the frequency of mar muos declined sharply in response to the spiraling cost of millet. However, at the same time, reliance on reciprocal work groups sharply increased.

8. Fragmentation here refers to diminution rather than the development of noncontiguous holdings (cf. King and Burton 1982). Cultivation of noncontiguous fields was unusual at the time of this study, although there were indications that it might be increasing with population density.

9. This is based on more reliable data than the slightly different production figures reported in Stone 1991a (p. 347).

10. Mar muos size is difficult to measure because the number of workers varies during the mar muos. The first mar muos I attended began with 21 workers, of whom 10 were working in the field while the rest were working their way through the muos (millet beer) prepared for the occasion. The size of the group fluctuated considerably over the next 3.5 hours, reaching 47 before I left.

11. The exception is the large type of mar muos called *lang long* ("the range of the chief"), in which every household in the ungwa is required to participate.

Chapter 8. Intensification, Dispersal, and Agglomeration

1. Pursuant to his concern with hierarchies of central places, Christaller's main discussion was on nested hierarchies within the hexagonal lattice. But because his model is based purely on the variables of distance and frequency of access, the hexagonal pattern fits my discussion of nonhierarchical movement in agriculture.

2. Travel cost where 20% of a constant number of trips are to a farm *x* farm widths away is calculated by adding 80% of the average intrafarm distance to 20% of the average interfarm distance. Average intrafarm distance is the average distance from the farm center to all points on a 0.5 unit grid on that farm; average interfarm distance is the average distance to the points on a farm that is a specified farm interval away, computed by measuring distances from the center of the "ego" farm to all points on a 0.5 unit grid on the farm in question.

3. The sample includes the farms of the seven labor study households and four others selected opportunistically. I would have preferred a sample that was much larger and random, but one of the main motivations for the mapping project was in recording land-use patterns, and there is a fairly narrow window near the middle of the agricultural season when this can be recorded reliably.

4. Three of the remaining four farms were recently fragmented, which restricts the farmer's ability to define the farm shape he wishes, and one had a large area removed against the owner's wishes.

5. On these farms, compound location is pulled simultaneously toward the farm center for economy of intrafarm movement and toward the road for econ-

omy of interfarm movement. Compounds are typically located 50–75 m off the road.

6. Adults are individuals over 14 years of age. There are now proportionately more children in Bong, and the total population has dropped only 20.4%.

7. The proportion of Muslims is highly sensitive to how much censusing we did in towns, and because the sample was not stratified to ensure representation of town dwellers, the estimate is unreliable. I do believe the relative proportions of traditionalist, Catholic, and Protestant to be reliable.

8. The attraction to other compounds is closely related to the attraction to roads and paths, which is taken up in chapter 11.

Chapter 9. Agricultural Movement

1. The exception is those few households that happen to be situated near the middle of roughly circular ungwa, for whom ungwa boundaries would be similar to a concentric work catchment.

2. Because this part of the analysis involves total numbers of compounds within given radii of the sample farms, it had to be restricted to Ungwa Kofyar. As figure 9.1 shows, the compounds of the Kwallala labor households are located near the edge of the mapped area.

3. It is also possible to isolate variation in travel patterns at the level of the individual household, as I have done previously (Stone 1991a).

Chapter 10. Ethnicity and Settlement

1. They called their portion of Wunze Yong, after the part of Kofyar village where the eponymous ancestor lived.

2. The southern half of KDG Koegoen, which is relatively swampy and was consequently settled much later than the northern half, is sometimes called Dangmel.

3. I use the two nearest neighbors rather than one nearest neighbor because a settler may be strongly attracted to a farmstead that turns out not to be its closest neighbor. Even if a farmer picks a location next to farmer X or is given a portion of farmer X's land, his compound may be erected closer to farmer Y's compound.

4. Fogol's value is low partly because it is socially close to both the Merniang sargwat (with which it is classified) and the Gankogom sargwat (with which it sometimes fought); several Fogol farmsteads on the frontier have no Merniang neighbors but do have Gankogom neighbors.

5. The great majority of agricultural movement is done on foot, although a few Kofyar are beginning to ride motorcycles to mar muos. Pedestrian travel times for 700 and 2000 m are roughly 15 and 45 minutes.

Chapter 11. Settlement and the Physical Landscape

1. In the homeland, water was transported in a ceramic vessel called a *kalang*, a wide-mouthed jar approximately 50 cm high with a flat base. Leaves placed on the surface of the water reduced spillage as the kalang was carried on the head. The kalang is still used in hill communities, where domestic water usually comes from springs. The kalang was used on the early frontier, but there it has been gradually replaced by large metal basins, also carried on the head. Water is fetched mostly by women and children.

2. The declining attraction to water has been documented among various other groups such as the Iban (Freeman 1955), the Tiv (Bohannan 1954a), and Irish farmers (Johnson 1958:554n). Archaeologically, the issue has been addressed by researchers working in the Maya Lowlands. Puleston and Puleston (1971) proposed a model in which swidden farmers settled first near rivers and later moved away from them, and Voorhies (1982:68–69) marshals data on early Maya settlement in support of this. Bogucki (1987:5) provides an example from an LBK settlement in northern Europe. In Germany, place name analysis reveals the tendency for early settlers to follow watercourses, with later villages being founded at greater distances from the water (Eidt 1976).

3. This analysis does not count instances of farms being abandoned for other farms in the ungwa, but this happened rarely. The analysis also does not discriminate between 1984 abandonments and farmers who had not left the ungwa by 1984, but there were few 1984 abandonments reported.

4. See G. D. Stone 1988 (pp. 206–215) for an analysis that includes settlement episodes where the respondent was not the household head. This can be an important difference, because a man who moved into an ungwa as a member of another man's household does not necessarily acquire his own farm there; his departure from the ungwa may signal a household fission rather than a case of abandonment.

5. I am also suspicious of some of the supposed abandonments of sand zone farms. Of the 20 cases of sand zone abandonments, 5 are from Hanyar Kwari in the 1970s and 1980s. The problem is that the terms *Kwari* and *Hanyar Kwari* are used inconsistently, and abandonments of farms on the clay soils toward Kwari (which were common during that period) may have been recorded as having occurred on the sandy soils of Hanyar Kwari. It is likely that the abandonment rate is even lower than 5%.

6. The upturn in sand zone site choice in the 1970s is from cases of farm fragmentation—owners of farms in the prime area turning part of their land over to another farmer, often a son who wants to establish his own household on the family farm. Fragmentation also occurs when an aged household head, whose domestic labor force and land needs have shrunk, cedes part of his land to a friend.

7. The effect of the census sample on the abandonment profile is most obvi-

ous in this zone. If my coworkers and I had censused in Bakin Ciawa, there would obviously have been more nonabandonment episodes figured into the Kwande pattern; but there would have also been many cases of abandonment, because many or most of the Bakin Ciawa farmers moved there from other Kwande-area farms. I believe that the pattern shown in figure 11.6 is in fact representative of the Kwande area, but I cannot be sure because of limited censusing there.

8. At the same time, soil bulk density problems caused drainage problems in residential courtyards. In Ungwa Kofyar alone I observed four compounds being rebuilt because of standing water in courtyard activity areas. These moves were intended not to bring the farmer closer to a new part of the farm, as in the case of Boyi of Ungwa Goewan, but to provide a fresh surface for extramural activities such as crop processing. Some of the farmers avoided moving by conducting maintenance work on their courtyards. One man in Ungwa Kofyar was demolishing an abandoned hut and using the adobe to repave the part of his courtyard used as a threshing floor. The repair was not only to fill in what had become a concave surface, but to provide a firmer (and drier) substrate for pounding grain.

Whereas drainage problems cause compound shifting in Ungwa Kofyar, they contribute to compound abandonment in poorer locations like Dangka. In Dangka, before a farmer would shift his compound because he wanted a better courtyard and threshing floor, he would probably seek a farm in a new ungwa because his yam yields were beginning to suffer.

9. In 1984, a restaurant meal in Namu of starch (such as pounded yam) and sauce cost ₦1, a price that doubled if a piece of chicken or goat was included. The farm diet, with cowpeas and occasionally fresh pork or chicken, offers ample protein, and many see the purchase of meat in restaurants as frivolous.

10. This attraction of the residence to the road is one reason agricultural intensity tends to be highest along roads; the area closest to the compound is the most intensively farmed.

Chapter 12. Agrarian Ecology and Culture

1. There will always be scholars of human behavior who see narratives as incomplete until they have been subsumed under nomothetic principles, and others who see this as doing damage to the uniqueness, complexity, and historical context of the events in question. In anthropology this dialectic has played out repeatedly, from Boas and the evolutionists down to Hodder and Binford, and it runs as well through work on agrarian settlements, as in Grossman 1971 contra Hudson 1969 and Jordan and Kaups 1989 contra Green 1979, 1980.

2. Compare Tiv market participation (Eyoh 1992:30) with that of the Kofyar (Netting et al. 1993).

3. The pattern of ephemeral settlement characteristic of the Tiv in the south-

ern Namu bush was not characteristic of all Tiv pioneers. A few compact villages with populations running into the hundreds, such as Tarkumburu (near Langkaku) and Asumeku (near Sabon Gida), were established before the Kofyar *entrada* and were still there in the 1990s.

4. In contrast, the swidden cultivation that cleared lands in the Amazon was followed by the even more extensive system of cattle grazing.

References

Abler, Ronald, J. S. Adams, and P. Gould
1977 *Spatial Organization: The Geographer's View of the World*. Prentice-Hall, London.

Adams, E. Charles
1979 Cold Air Drainage and Length of Growing Season in the Hopi Mesas Area. *Kiva* 44:285–297.

Adams, Robert McC.
1974 Anthropological Perspectives on Trade. *Current Anthropology* 15: 239–258.

Adamu, Mahdi
1978 *The Hausa Factor in West African History*. Ahmadu Bello Press, Zaria.

Agi, John Ola
1982 The Goemai and Their Neighbors: An Historical Analysis. In *Studies in the History of Plateau State, Nigeria*, edited by Elizabeth Isichei, pp. 98–107. Macmillan, London.

Ahlstrom, Richard V. N., Jeffrey S. Dean, and William J. Robinson
1991 Evaluating Tree-Ring Interpretations at Walpi Pueblo, Arizona. *American Antiquity* 56:628–644.

Aiken, Charles S.
1985 New Settlement Pattern of Rural Blacks in the American South. *Geographical Review* 75:383–404.

Ajakaiye, D. E., and A. E. Scheidegger
1989 Joints on the Jos Plateau. *Journal of African Earth Sciences* 9:725–727.

Aldskogius, Hans
1969 *Modelling the Evolution of Settlement Patterns: Two Studies of Vacation House Settlement*. Geografiska Regionstudier 6, Uppsala University, Sweden.

Amanor, Kojo S.

1994 *The New Frontier: Farmers' Response to Land Degradation*. Zed Books, London.

Ambler, Charles

1988 *Kenyan Communities in the Age of Imperialism*. Yale University Press, New Haven.

Ames, C. G.

1934 *Gazetteer of the Plateau Province*. Jos Native Administration, Niger Press, Jos.

Anthony, Kenneth R. M., B. F. Johnston, W. O. Jones, and V. C. Uchendu

1979 *Agricultural Change in Tropical Africa*. Cornell University Press, Ithaca.

Appleton, John, and D. Symes

1986 Family Goals and Kinship Strategies in a Capitalist Farming Society. Paper at the 13th European Congress for Rural Sociology, Braga, Portugal.

Ashmore, Wendy

1981 Some Issues of Method and Theory in Lowland Maya Settlement Archaeology. In *Lowland Maya Settlement Patterns*, edited by Wendy Ashmore, pp. 37–69. University of New Mexico Press, Albuquerque.

Austin, David

1985 Doubts about Morphogenesis. *Journal of Historical Geography* 11: 201–209.

Baker, A.R.H.

1969 The Geography of Rural Settlements. In *Trends in Geography—An Introduction*, edited by R. U. Cooke and J. H. Johnson, pp. 123–132. Pergamon Press, Oxford, U.K.

Baltensperger, Bradley H.

1993 Larger and Fewer Farms: Patterns and Causes of Farm Enlargement on the Central Great Plains, 1930–1978. *Journal of Historical Geography* 19:299–313.

Barth, Fredrik

1956 Ecological Relationships of Ethnic Groups in Swat, North Pakistan. *American Anthropologist* 58:1079–1089.

Basehart, Harry W.

1973 Cultivation Intensity, Settlement Patterns, and Homestead Farms among the Matengo of Tanzania. *Ethnology* 12:57–74.

Baum, E.

1968 Land Use in the Kilombero Valley. In *Smallholder Farmer and Smallholder Development in Tanzania*, edited by Hans Ruthenberg, pp. 23–50. IFO Institut, Munich.

Bawden, M. G., and J. A. Jones

1972 The Physiography of the Benue Valley: Interim Report on the Land Resources of Central Nigeria. Ministry of Overseas Development, Land Resources Development Centre, Miscellaneous Report 128. Surbiton, Surrey, England.

Bayliss-Smith, T. P.

1982 *The Ecology of Agricultural Systems.* Cambridge University Press, Cambridge.

Beckerman, Stephen

1987 Swidden in Amazonia and the Amazon Rim. In *Comparative Farming Systems,* edited by B. L. Turner II and S. B. Brush, pp. 55–94. Guilford Press, New York.

Benkhelil, J.

1989 The Origin and Evolution of the Cretaceous Benue Trough (Nigeria). *Journal of African Earth Sciences* 8:251–282.

Bennett, J. G., I. D. Hill, W. J. Howard, A. A. Hutcheon, L. J. Rackham, A. W. Wood

1976 Land Resources of Central Nigeria: Landforms, Soils and Vegetation of the Benue Valley. Ministry of Overseas Development, Land Resources Development Centre, Land Resource Report 7. Surbiton, Surrey, U.K.

Bennett, John W.

1967 *Hutterian Brethren: The Agricultural Economy and Social Organization of a Communal People.* Stanford University Press, Stanford.

1969 *Northern Plainsmen: Adaptive Strategy and Agrarian Life.* Aldine, Chicago.

Bentley, Jeffrey W.

1987 Economic and Ecological Approaches to Land Fragmentation: In Defense of a Much-Maligned Phenomenon. *Annual Reviews in Anthropology* 16:31–67.

Berry, Brian J. L., and H. G. Barnum

1962 Aggregate Relations and Elemental Components of Central Place Systems. *Journal of Regional Science* 4:35–68.

Berry, Sara

1993 *No Condition Is Permanent: The Social Dynamics of Agrarian Change in Sub-Saharan Africa.* University of Wisconsin Press, Madison.

Bilsborrow, Richard E., and Martha Geores

1994 Population Change and Agricultural Intensification in Developing Countries. In *Population and Environment: Rethinking the Debate,* edited by Lourdes Arizpe, M. Priscilla Stone, and David C. Major, pp. 171–207. Westview Press, Boulder.

Binford, Lewis R.

1964 A Consideration of Archaeological Research Design. *American Antiquity* 29:425–441.

1968a Archaeological Perspectives. In *New Perspectives in Archaeology*, edited by S. R. Binford and L. R. Binford, pp. 5–32. Aldine, Chicago.

1968b Post-Pleistocene Adaptations. In *New Perspectives in Archeology*, edited by Sally R. Binford and L. R. Binford, pp. 313–341. Aldine Publishing, New York.

1971 Spatial Organization of Agriculture in Some North Indian Villages, Part I. *Transactions of the Institute of British Geographers* 52:1–40.

1980 Willow Smoke and Dogs' Tails: Hunter-Gatherer Settlement Systems and Archaeological Site Formation. *American Antiquity* 45: 4–20.

1990 Mobility, Housing, and Environment: A Comparative Study. *Journal of Anthropological Research* 46:119–152.

Blaikie, Piers, and Harold Brookfield

1987 *Land Degradation and Society*. Methuen, London.

Blok, Anton

1969 South Italian Agro-Towns. *Comparative Studies in Society and History* 11:121–135.

Blouet, Brian W.

1972 Factors Influencing the Evolution of Settlement Patterns. In *Man, Settlement and Urbanism*, edited by Peter J. Ucko, R. Tringham, and G. W. Dimbleby, pp. 3–15. Schenkman, Cambridge, Mass.

Boatright, Mody

1941 The Myth of Frontier Individualism. *Southwestern Social Science Quarterly* 22:12–23.

Bogucki, Peter

1987 The Establishment of Agrarian Communities on the North European Plain. *Current Anthropology* 28:1–24.

Bohannan, Paul

1954a *Tiv Farm and Settlement*. HMSO, London.

1954b The Migration and Expansion of the Tiv. *Africa* 24:2–16.

Bohland, James R.

1970 The Influence of Kinship Ties on the Settlement Pattern of Northeast Georgia. *Professional Geographer* 22:267–269.

Boserup, Ester

1965 *The Conditions of Agricultural Growth*. Aldine, New York.

1970 Present and Potential Food Production in Developing Countries. In *Geography and a Crowding World*, edited by W. Zelinsky, L. A. Kosinski, and R. M. Prothero, pp. 100–113. Oxford University Press, London.

1981 *Population and Technological Change: A Study of Long Term Trends.* University of Chicago Press, Chicago.

Bromley, Yu

1978 On the Typology of Ethnic Communities. In *Perspectives on Ethnicity,* edited by R. E. Holloman and S. A. Artuiunov, pp. 15–21. Mouton, The Hague.

Bronson, Bennet

1972 Farm Labor and the Evolution of Food Production. In *Population Growth: Anthropological Implications,* edited by Brian Spooner, pp. 190–218. MIT Press, Cambridge.

1975 The Earliest Farming: Demography as Cause and Consequence. In *Population, Ecology and Social Evolution,* edited by Steven Polgar, pp. 53–78. Mouton, The Hague.

Brookfield, Harold C.

1968 New Directions in the Study of Agricultural Systems in Tropical Areas. In *Evolution and Environment,* edited by Ellen T. Drake, pp. 413–439. Yale University Press, New Haven.

1972 Intensification and Disintensification in Pacific Agriculture. *Pacific Viewpoint* 13:30–41.

1984 Intensification Revisited. *Pacific Viewpoint* 25:15–44.

Brookfield, Harold C., and Paula Brown

1963 *Struggle for Land.* Oxford University Press, Melbourne.

Brown, James A., Robert E. Bell, and Don G. Wyckoff

1978 Caddoan Settlement Patterns in the Arkansas River Drainage. In *Mississippian Settlement Patterns,* edited by Bruce D. Smith, pp. 169–200. Academic Press, New York.

Brown, Paula, and Aaron Podolefsky

1976 Population Density, Agricultural Intensity, Land Tenure and Group Size in the New Guinea Highlands. *Ethnology* 15:211–238.

Bruce, Richard

1982 The Growth of Islam and Christianity: The Pyem Experience. In *Studies in the History of Plateau State, Nigeria,* edited by Elizabeth Isichei, pp. 224–241. Macmillan, London.

Brunger, Alan G.

1975 Early Settlement in Contrasting Areas of Peterborough County, Ontario. In *Perspectives on Landscape and Settlement in Nineteenth Century Ontario,* edited by J. David Wood, pp. 117–140. McClelland and Stewart, Toronto.

1982 Geographical Propinquity among Pre-Famine Catholic Irish Settlers in Upper Canada. *Journal of Historical Geography* 8:265–282.

Buchanan, K. M., and J. C. Pugh

1955 *Land and People in Nigeria.* University of London Press, London.

Bunge, William
1962 *Theoretical Geography*. C.W.K. Gleerup, Lund, Sweden.

Burghardt, Andrew F.
1959 The Location of River Towns in the Central Lowland of the United States. *Annals of the Association of American Geographers* 49:305–323.

Burnham, Philip
1980 Changing Agricultural and Pastoral Ecologies in the West African Savanna Region. In *Human Ecology in Savanna Environments*, edited by David R. Harris, pp. 147–170. Academic Press, London.

Butt, Audrey J.
1977 Land Use and Social Organization of Tropical Forest Peoples of the Guianas. In *Human Ecology in the Tropics*, edited by J. P. Garlick and R.W.J. Keay, pp. 1–17. Taylor and Francis, London.

Bylund, Erik
1960 Theoretical Considerations Regarding the Distribution of Settlement in Inner North Sweden. *Geografiska Annaler B* 42:225–231.

Carlstein, Tommy
1982 *Time Resources, Society and Ecology*, vol. 1, *Preindustrial Societies*. George Allen & Unwin, London.

Carlyle, William J.
1983 The Changing Family Farm on the Prairies. *Prairie Forum* 8(1):1–23.

Carneiro, Robert L.
1960 Slash-and-Burn Agriculture: A Closer Look at Its Implications for Settlement Patterns. In *Men and Cultures: Selected Papers of the 5th Intl. Congress of Anth. and Ethn. Sciences*, edited by Anthony F. C. Wallace, pp. 229–234. University of Pennsylvania Press, Philadelphia.

Chang, K. C.
1958 Study of the Neolithic Social Grouping: Examples from the New World. *American Anthropologist* 60:298–334.

Chanock, Martin
1985 *Law, Custom and Social Order*. Cambridge University Press, Cambridge.

Chapdelaine, Claude
1993 The Sedentarization of the Prehistoric Iroquoians: A Slow or Rapid Transformation? *Journal of Anthropological Archaeology* 12:173–209.

Chayanov, A. V.
1966 *The Theory of Peasant Economy*. Edited by D. Thorner, B. Kerblay, and R. Smith. Richard D. Irwin, Homewood, Ill. Originally published 1925.

Chibnik, Michael, and Wil de Jong
1989 Agricultural Labor Mobilization in Ribereño Communities of the Peruvian Amazon. *Ethnology* 28:75–95.

Chisholm, Michael

1962 *Rural Settlement and Land Use: An Essay in Location.* Hutchinson University Library, London.

1968 *Rural Settlement and Land Use: An Essay in Location.* 2d ed. Hutchinson University Library, London.

1979 *Rural Settlement and Land Use: An Essay in Location.* 3d ed. Hutchinson University Library, London.

Christaller, Walter

1966 *Central Places in Southern Germany.* Translated by C. W. Baskin. Prentice-Hall, Englewood Cliffs, N.J. Originally published 1933.

Clark, C., and M. R. Haswell

1967 *The Economics of Subsistence Agriculture.* Macmillan, London.

Cleave, J. H.

1974 *African Farmers: Labor Use in the Development of Smallholder Agriculture.* Praeger, New York.

Collier, G. A.

1975 Are Marginal Farmlands Marginal to Their Farmers? In *Formal Methods in Economic Anthropology,* edited by Stuart Plattner, pp. 149–158. American Anthropological Association, Special Publications no. 4. Washington, D.C.

Conant, Francis P.

1962 Contemporary Communities and Abandoned Settlement Sites. *Annals of the New York Academy of Sciences* 96:539–574.

Conelly, W. Thomas

1992 Agricultural Intensification in a Philippine Frontier Community: Impact on Labor Efficiency and Farm Diversity. *Human Ecology* 20:203–223.

Connah, Graham

1985 Agricultural Intensification and Sedentism in the Firki of N. E. Nigeria. In *Prehistoric Intensive Agriculture in the Tropics,* edited by I. S. Farrington, pp. 765–785. BAR International Series 232. London.

Cordell, Linda S.

1975 Predicting Site Abandonment at Weatherill Mesa. *Kiva* 40:189–202.

Cordell, Linda S., and Fred Plog

1979 Escaping the Confines of Normative Thought: A Reevaluation of Puebloan Prehistory. *American Antiquity* 44:405–429.

Cornell, Stephen

1988 The Transformation of Tribe: Organization and Self-Concept in Native American Ethnicities. *Ethnic and Racial Studies* 11:27–47.

Crumley, Carole L.

1979 Three Locational Models: An Epistemological Assessment for Anthropology and Archaeology. In *Advances in Archaeological Method and Theory*, vol. 3, edited by Michael B. Schiffer, pp. 141–173. Academic Press, New York.

Crumley, Carole L., editor
1994 *Historical Ecology: Cultural Knowledge and Changing Landscapes*. School of American Research Press, Santa Fe.

Dagum, Naanshep Fogotnaan
n.d. The Development of the People of Namu (Njak) with Special Reference to Migration and Intergroup Relations. B.A. History Thesis, University of Jos.

Damp, Jonathan E.
1984 Environmental Variation, Agriculture, and Settlement Processes in Coastal Ecuador (3300–1500 B.C.). *Current Anthropology* 25: 106–111.

Deetz, James
1968 The Inference of Residence and Descent Rules from Archeological Data. In *New Perspectives in Archeology*, edited by Sally R. Binford and L. R. Binford, pp. 41–48. Aldine Publishing, New York.

DeLisle, Davis
1982 Effects of Distance on Cropping Patterns Internal to the Farm. *Annals of the Association of American Geographers* 72:88–98.

Demangeon, Albert
1927 La géographie de l'habitat rural. *Annales de Géographie* 36:1–23. Reprinted as "The Origins and Causes of Settlement Types" in *Readings in Cultural Geography* (1962), edited by P. L. Wagner and M. W. Mikesell, pp. 506–516. University of Chicago Press, Chicago.

De Montmollin, Olivier
1989 Land Tenure and Politics in the Late/Terminal Classic Rosario Valley. *Journal of Anthropological Research* 45:293–314.

Denevan, William M.
1992 Stone vs Metal Axes: The Ambiguity of Shifting Cultivation in Prehistoric Amazonia. *Journal of the Steward Anthropological Society* 20(1):153–165.

Denevan, W. M., and B. L. Turner II
1974 Forms, Functions and Associations of Raised Fields in the Old World Tropics. *Journal of Tropical Geography* 39:1–24.

Dickenson, Robert E.
1949 Rural Settlement in the German Lands. *Annals of the Association of American Geographers* 34:239–263.

Donham, Donald
1990 *History, Power and Ideology: Central Issues in Marxism and Anthropology*. Cambridge University Press, Cambridge.

Donkin, R. A.

1979 *Agricultural Terracing in the Aboriginal New World*. Viking Fund Publications in Anthropology 56. University of Arizona Press, Tucson.

Doolittle, William E.

1980 Aboriginal Agricultural Development in the Valley of Sonora, Mexico. *Geographical Review* 70:328–342.

1990 *Canal Irrigation in Prehistoric Mexico: The Sequence of Technological Change*. University of Texas Press, Austin.

Drucker, C. B.

1977 To Inherit the Land: Descent and Decision in Northern Luzon. *Ethnology* 21:1–20.

Dunham, Peter

1989 Hudson's Model of Settlement Location and the Classic Expansion of Maya Civilization. Paper at the meeting of the American Anthropological Association, Washington, D.C.

du Toit, Brian M.

1978 Introduction. In *Ethnicity in Modern Africa*, edited by Brian M. du Toit, pp. 1–16. Westview Press, Boulder.

Earle, Timothy K.

1976 A Nearest-Neighbor Analysis of Two Formative Settlement Systems. In *The Early Mesoamerican Village*, edited by Kent V. Flannery, pp. 196–223. Academic Press, New York.

1980 A Model of Subsistence Change. In *Modeling Change in Prehistoric Subsistence Economies*, edited by Timothy K. Earle and A. L. Christenson, pp. 1–29. Academic Press, New York.

Easterlin, Richard A.

1976 Population Change and Farm Settlement in the Northern United States. *Journal of Economic History* 36:45–75.

Eder, James F.

1982 *Who Shall Succeed? Agricultural Development and Social Inequality on a Philippine Frontier*. Cambridge University Press, Cambridge.

1991 Agricultural Intensification and Labor Productivity in a Philippine Vegetable Garden Community: A Longitudinal Study. *Human Organization* 50:245–255.

Effland, R. W.

1979 A Study of Prehistoric Spatial Behavior: Long House Valley, Arizona. Ph.D. diss., Department of Anthropology, Arizona State University, Tempe.

Eidt, Robert C.

1971 *Pioneer Settlement in Northeast Argentina*. University of Wisconsin Press, Madison.

1976 Some Observations on the Role of Geography and History in the Development of a Methodology for the Analysis of Settlements. In

Geographic Dimensions of Rural Settlement, edited by R. L. Singh, K. N. Singh, and R.P.B. Singh, pp. 7–18. National Geographical Society of India, Varanasi.

1984 *Advances in Abandoned Settlement Analysis: Application to Prehistoric Anthrosols in Columbia, South America.* Center for Latin America, University of Wisconsin, Milwaukee.

Ensminger, Jean

1992 *Making a Market: The Institutional Transformation of an African Society.* Cambridge University Press, New York.

Erasmus, Charles J.

1956 Culture Structure and Process: The Occurrence and Disappearance of Reciprocal Farm Labor. *Southwestern Journal of Anthropology* 12:444–469.

Erickson, Clark L.

1993 Social Organization of Prehispanic Raised Field Agriculture in the Lake Titicaca Basin. In "Economic Aspects of Water Management in the Prehispanic New World." *Research in Economic Anthropology* Supplement 7:369–426.

Evans, Susan T.

1980 Spatial Analysis of Basin of Mexico Settlement: Problems with the Use of the Central Place Model. *American Antiquity* 45:866–875.

Evans, Susan T., and Peter Gould

1982 Settlement Models in Archaeology. *Journal of Anthropological Archaeology* 1:275–304.

Eyoh, Dickson

1992 Structures of Intermediation and Change in African Agriculture: A Nigerian Case Study. *African Studies Review* 35:17–39.

Farriss, Nancy M.

1978 Nucleation vs. Dispersal: The Dynamics of Population Movement in Colonial Yucatan. *Hispanic American Historical Review* 5:187–216.

Feinman, Gary, Richard Blanton, and Stephen Kowalewski

1984 Market System Development in the Prehistoric Valley of Oaxaca, Mexico. In *Trade and Exchange in Early Mesoamerica,* edited by Kenneth G. Hirth, pp. 157–178. University of New Mexico Press, Albuquerque.

Feinman, Gary M., Linda M. Nicholas, and William D. Middleton

1992 Archaeology in 1992: A Perspective on the Discipline from the Society for American Archaeology Annual Program. *American Antiquity* 57:448–458.

Findlay, R. L.

1945 The Dimmuk and Their Neighbors. *Farm and Forest* 6:137–145.

Fitzpatrick, J.F.J.

1910 Some Notes on the Kwolla District and Its Tribes. *Journal of the African Society* 10:16–52.

Flannery, Kent V.

1972a The Origins of the Village as a Settlement Type in Mesoamerica and the Near East: A Comparative Study. In *Man, Settlement and Urbanism*, edited by Peter J. Ucko, R. Tringham, and G. W. Dimbleby, pp. 23–53. Schenkman, Cambridge, Mass.

1972b The Cultural Evolution of Civilizations. *Annual Review of Ecology and Systematics* 3:399–426.

1976a Linear Stream Patterns and Riverside Settlement Rules. In *The Early Mesoamerican Village*, edited by Kent V. Flannery, pp. 173–180. Academic Press, New York.

1976b Evolution of Complex Settlement Systems. In *The Early Mesoamerican Village*, edited by Kent V. Flannery, pp. 162–173. Academic Press, New York.

1986 A Visit to the Master. In *Guila Naquitz: Archaic Foraging and Early Agriculture in Oaxaca, Mexico*, edited by Kent V. Flannery, pp. 511–519. Academic Press, Orlando.

Flannery, Kent V., A.V.T. Kirby, M. J. Kirby, and A. W. Williams, Jr.

1967 Farming Systems and Political Growth in Ancient Oaxaca. *Science* 158:445–454.

Flannery, Kent V., J. Marcus, and R. Reynolds

1989 *The Flocks of the Wamani: A Study of Llama Herders on the Punas of Ayacucho, Peru*. Academic Press, San Diego.

Flora, Jan L., and John M. Stitz

1985 Ethnicity, Persistence, and Capitalization of Agriculture in the Great Plains during the Settlement Period: Wheat Production and Risk Avoidance. *Rural Sociology* 50:341–360.

Folorunso, C. A., and S. O. Ogundele

1993 Agriculture and Settlement among the Tiv of Nigeria: Some Ethnoarchaeological Observations. In *The Archaeology of Africa: Food, Metal and Towns*, edited by Thurstan Shaw, Paul Sinclair, Bassey Andah, and Alex Okpoko, pp. 274–288. Routledge, London.

Foster, Gary, R. Hummel, and R. Whittenbarger

1987 Ethnic Echoes through 100 Years of Midwestern Agriculture. *Rural Sociology* 52:365–378.

Freeman, Derek

1955 *Iban Agriculture: A Report on the Shifting Cultivation of Hill Rice by the Iban of Sarawak*. HMSO, London.

Fremantle, J. M., editor

1922 *Gazetteer of Muri Province*. Waterlow & Sons, London.

Fried, Morton H.
1975 *The Notion of Tribe*. Cummings, Menlo Park, Calif.

Friesen, Richard J.
1977 Saskatchewan Mennonite Settlements: The Modification of an Old World Settlement Pattern. *Canadian Ethnic Studies* 9:72–90.

Gade, Daniel W., and Mario Escobar
1982 Village Settlement and the Colonial Legacy in Southern Peru. *Geographical Review* 72:430–449.

Gall, P. L., and A. Saxe
1977 The Ecological Evolution of Culture: The State as Predator in Succession Theory. In *Exchange Systems in Prehistory*, edited by T. Earle and J. Ericson, pp. 255–267. Academic Press, New York.

Galt, Anthony H.
1979 Exploring the Cultural Ecology of Field Fragmentation and Scattering on the Island of Pantelleria, Italy. *Journal of Anthropological Research* 35:93–108.
1991 *Far from the Church Bells: Settlement and Society in an Apulian Town*. Cambridge University Press, Cambridge.

Gandonu, Ajato
1978 Nigeria's 250 Ethnic Groups: Realities and Assumptions. In *Perspectives on Ethnicity*, edited by R. E. Holloman and S. A. Artuiunov, pp. 243–279. Mouton, The Hague.

Gjerde, Jon
1979 The Effect of Community on Migration: Three Minnesota Townships 1885–1905. *Journal of Historic Geography* 5:403–422.

Glassan, B. G.
1927 The Mirriam Tribe. Jos Provincial Files 2/27, 1399. National Archives, Kaduna.

Glassow, Michael A.
1977 Population Aggregation and Systemic Change: Examples from the American Southwest. In *Explanation of Prehistoric Change*, edited by James N. Hill, pp. 185–214. University of New Mexico Press, Albuquerque.

Gleave, Michael B.
1963 Hill Settlements and Their Abandonment in Western Yorubaland. *Africa* 33:343–352.
1965 The Changing Frontiers of Settlement in the Uplands of Northern Nigeria. *Nigerian Geographical Journal* 8:127–141.
1966 Hill Settlements and Their Abandonment in Tropical Africa. *Transactions of the Institute of British Geographers* 40:39–49.

Gleave, Michael B., and H. P. White
1969 Population Density and Agricultural Systems in Africa. In *Environ-*

ment and Land Use in Africa, edited by M. F. Thomas and G. W. Whittington, pp. 273–300. Methuen, London.

Goldman, Abe

1993 Population Growth and Agricultural Change in Imo State, Southeastern Nigeria. In *Population Growth and Agricultural Change in Africa,* edited by B. L. Turner II, Goran Hyden, and Robert W. Kates, pp. 250–301. University of Florida Press, Gainesville.

Goldschmidt, Walter

1970 Game Theory, Cultural Values and the Brideprice in Africa. In *Game Theory in the Behavioral Sciences,* edited by I. R. Buchler and H. G. Nutini, pp. 61–74. University of Pittsburgh Press, Pittsburgh.

Graves, Michael W., William A. Longacre, and Sally J. Holbrook

1982 Aggregation and Abandonment at Grasshopper Pueblo, Arizona. *Journal of Field Archaeology* 9:193–206.

Green, Stanton W.

1979 The Agricultural Colonization of Temperate Forest Habitats: An Ecological Model. In *The Frontier: Comparative Studies,* vol. 2, edited by William W. Savage and S. I. Thompson, pp. 69–103. University of Oklahoma Press, Norman.

1980 Toward a General Model of Agricultural Systems. In *Advances in Archaeological Method and Theory,* vol. 3, edited by Michael B. Schiffer, pp. 311–355. Academic Press, New York.

Grigg, David B.

1979 Ester Boserup's Theory of Agrarian Change: A Critical Review. *Progress in Human Geography* 3:64–84.

1980 *Population Growth and Agrarian Change.* Cambridge University Press, Cambridge.

Gross, Daniel R.

1975 Protein Capture and Cultural Development in the Amazon. *American Anthropologist* 77:526–549.

1983 Village Movement in Relation to Resources in Amazonia. In *Adaptive Responses of Native Amazonians,* edited by Raymond B. Hames and William T. Vickers, pp. 429–449. Academic Press, New York.

Grossman, David

1971 Do We Have a Theory for Settlement Geography?—The Case of Iboland. *Professional Geographer* 3:197–203.

Grove, A. T.

1952 *Land Use and Soil Conservation on the Jos Plateau.* Geological Survey of Nigeria, Bulletin 22. Federal Republic of Nigeria, Lagos.

1961 Population Densities and Agriculture in Northern Nigeria. In *Essays on African Population,* edited by K. M. Barbour and R. M. Prothero, pp. 115–136. Routledge & Kegan Paul, London.

Guillet, David
1980 Reciprocal Labor and Peripheral Capitalism in the Central Andes. *Ethnology* 19:151–167.

Guyer, Jane I.
1992 Small Change: Individual Farm Work and Collective Life in a Western Nigerian Savanna Town, 1969–88. *Africa* 62:465–489.
n. d. African Farmers and Their Environment in Long Term Perspective. Conference proposal to the National Science Foundation, 1993.

Guyer, Jane I., and Eric F. Lambin
1993 Land Use in an Urban Hinterland: Ethnography and Remote Sensing in the Study of African Intensification. *American Anthropologist* 95:839–859.

Haberland, Wolfgang
1983 To Quench the Thirst: Water and Settlement in Central America and Beyond. In *Prehistoric Settlement Patterns: Essays in Honor of Gordon R. Willey*, edited by Evon Z. Vogt and Richard M. Leventhal, pp. 79–88. University of New Mexico Press, Albuquerque.

Haggett, Peter
1965 *Locational Analysis in Human Geography*. Edward Arnould, London.
1972 *Geography—A Modern Synthesis*. Harper and Row, New York.

Haining, Robert
1982 Describing and Modeling Rural Settlement Maps. *Annals of the Association of American Geographers* 72:211–223.

Hames, Raymond, and William Vickers
1982 Optimal Diet Breadth Theory as a Model to Explain Variability in Amazonian Hunting. *American Ethnologist* 9:358–378.

Hamond, Fred
1981 The Colonisation of Europe: The Analysis of Settlement Process. In *Pattern of the Past: Studies in Honor of David Clarke*, edited by Ian Hodder, G. Issac, and N. Hammond, pp. 211–248. Cambridge University Press, Cambridge.

Hanks, L. M.
1972 *Rice and Man: Agricultural Ecology in Southeast Asia*. Aldine, Chicago.

Hansen, B.
1979 Colonial Economic Development with Unlimited Supply of Land: A Ricardian Case. *Economic Development and Cultural Change* 27: 611–627.

Hantman, Jeff
1978 Models of the Explanation of Changing Settlement on the Little Colorado River. *Arizona State University Anthropological Research Paper* 13:169–187.

Hard, Robert J., and William L. Merrill
1992 Mobile Agriculturalists and the Emergence of Sedentism: Perspectives from Northern Mexico. *American Anthropologist* 94:601–620.

Hart, Keith
1982 *The Political Economy of West African Agriculture*. Cambridge University Press, Cambridge.

Haynes, Kingsley E., and W. T. Enders
1975 Distance, Direction and Entropy in the Evolution of a Settlement Pattern. *Economic Geography* 51:357–365.

Hecht, Susanna B.
1981 Deforestation in the Amazon Basin: Magnitude, Dynamics and Soil Resource Effects. *Studies in Third World Societies* 13:61–100.

Hecht, Susanna, and Alexander Cockburn
1989 *The Fate of the Forest: Developers, Destroyers, and Defenders of the Amazon*. Verso, London.

Heider, Karl
1970 *The Dugum Dani*. Viking Fund Publication in Anthropology 49. University of Arizona Press, Tucson.

Henshall, J. D.
1967 Models of Agricultural Activity. In *Models in Geography*, edited by R. J. Chorley and P. Haggett, pp. 425–458. Methuen, London.

Heston, Alan, and D. Kumar
1983 The Persistence of Land Fragmentation in Peasant Agriculture: An Analysis of South Asian Cases. *Explanations in Economic History* 20:199–220.

Higgs, E. S., and C. Vita-Finzi
1972 Prehistoric Economies: A Territorial Approach. In *Papers in Economic Prehistory*, edited by E. S. Higgs, pp. 27–36. Cambridge University Press, Cambridge.

Higgs, E. S., C. Vita-Finzi, D. R. Harriss, and A. E. Fagg
1967 The Climate, Environment and Industries of Stone Age Greece, part 3. *Proceedings of the Prehistoric Society* 33:1–29.

Hill, I. D., editor
1979 *Land Resources of Central Nigeria, Agricultural Development Possibilities*, vol. 4, *The Benue Valley*. Ministry of Overseas Development, Land Resources Development Centre, Land Resource Study 29. Surbiton, Surrey, England.

Hill, James N.
1970 *Broken K Pueblo: Prehistoric Social Organization in the American Southwest*. Anthropological Papers of the University of Arizona no. 18. Tucson.

Hill, Polly
1977 *Population, Prosperity and Poverty: Rural Kano, 1900 and 1970*. Cambridge University Press, Cambridge.

Hodder, Ian
1977 Some New Directions in the Spatial Analysis of Archaeological Data at the Regional Scale. In *Spatial Archaeology*, edited by David L. Clarke, pp. 223–351. Academic Press, New York.
1981 Society, Economy and Culture: An Ethnographic Case Study amongst the Lozi. In *Patterns of the Past: Studies in Honor of David Clarke*, edited by I. Hodder, G. Issac, and N. Hammond, pp. 67–96. Cambridge University Press, Cambridge.
1982 *Symbols in Action: Ethnoarchaeological Studies of Material Culture*. Cambridge University Press, Cambridge.

Hodder, Ian, and Mark Hassall
1971 The Non-Random Spacing of Romano-British Walled Towns. *Man* 6:391–407.

Hodder, Ian, and C. Orton
1976 *Spatial Analysis in Archaeology*. Cambridge University Press, Cambridge.

Hogben, S. J.
1930 *The Muhammadan Emirates of Nigeria*. Oxford University Press, London.

Hogben, S. J., and A.H.M. Kirk-Greene
1966 *The Emirates of Northern Nigeria*. Oxford University Press, London.

Holzall, Vince, and Glenn Davis Stone
1990 Dispersed Settlements, Invisible Villages. Paper at the Meeting of the Society for American Archaeology, Las Vegas.

Hornby, William F., and Malvyn Jones
1991 *An Introduction to Settlement Geography*. Cambridge University Press, Cambridge.

Horvath, R. J
1969 Von Thunen's Isolated State and the Area around Addis Ababa, Ethiopia. *Annals of the Association of American Geographers* 59: 308–323.

Huang, Philip
1990 *The Peasant Family and Rural Development in the Yangzi Delta, 1350–1988*. Stanford University Press, Stanford.

Hudson, John C.
1969 A Location Theory for Rural Settlement. *Annals of the Association of American Geographers* 59:365–381.
1977 Process Models of Settlement Patterns. In *Man, Culture and Settlement*, edited by Robert C. Eidt, K. N. Singh, and R.P.B. Singh, pp. 238–247. Kalyani, New Delhi.

Hunter, John C.
1963 Cocoa Migration and Patterns of Land Ownership in the Densu Valley near Suhum, Ghana. *Transactions of the Institute of British Geographers* 33:61–87.
1967 The Social Roots of Dispersed Settlement in Northern Ghana. *Annals of the Association of American Geographers* 57:338–349.

Hyden, Goran
1980 *Beyond Ujamaa in Tanzania: Underdevelopment and an Uncaptured Peasantry*. University of California Press, Berkeley.

Isard, Walter
1960 *Methods of Regional Analysis*. MIT Press, Cambridge.

Jackson, Richard
1972 A Vicious Circle?—The Consequences of von Thünen in Tropical Africa. *Area* 4:258–261.

Jarman, M. R., G. N. Bailey, and H. N. Jarman
1982 *Early European Agriculture: Its Foundations and Development*. Cambridge University Press, Cambridge.

Jochim, Michael A.
1991 Archaeology as Long-Term Ethnography. *American Anthropologist* 93:308–321.

JOHLT (Jos Oral History and Literature Texts)
1981 Volume 1: Mwahavul, Ngas, Mupun, Njak. Department of History, University of Jos.

Johnson, Gregory A.
1972 A Test of the Utility of Central Place in Archaeology. In *Man, Settlement and Urbanism*, edited by Peter J. Ucko, R. Tringham, and G. W. Dimbleby, pp. 769–785. Duckworth, London.
1977 Aspects of Regional Analysis in Archaeology. *Annual Reviews in Anthropology* 6:479–508.

Johnson, Hildegard B.
1941 Distribution of German Pioneer Population in Minnesota. *Rural Sociology* 6:16–34.
1945 Factors Influencing the Distribution of the German Pioneer Population in Minnesota. *Agricultural History* 19:39–57.
1951 The Location of German Immigrants in the Middle West. *Annals of the Association of American Geographers* 41:1–41.

Johnson, James H.
1958 Studies of Irish Rural Settlement. *Geographical Review* 48:554–566.

Jolley, Clifford J., and Fred Plog
1987 *Physical Anthropology and Archaeology*. 4th ed. Alfred A. Knopf, New York.

Jones, G.C.I.
1945 Agriculture and Ibo Village Planning. *Farm and Forest* 6:9–15.

Jordan, Terry
1966 On the Nature of Settlement Geography. *Professional Geographer* 28:26–27.
1976 Abandonment of Farm-Village Tradition in New-Land Settlement: The Example from Anglo-America. In *Geographic Dimensions of Rural Settlement*, edited by R. L. Singh, K. N. Singh, and R.P.B. Singh, pp. 84–88. National Geographical Society of India, Varanasi.

Jordan, Terry G., and Matti Kaups
1989 *The American Backwoods Frontier: An Ethnic and Ecological Interpretation*. Johns Hopkins University Press, Baltimore.

Josprof (Jos Provincial Files)
n.d. National Archives, Kaduna, Nigeria. File 1197, "Dimmuk Sub-Tribe Disturbance, Shendam Division."

Kalipeni, Ezekiel
1994 Population Growth and Environmental Degradation in Malawi. In *Population Growth and Environmental Degradation in Southern Africa*, edited by Ezekiel Kalipeni, pp. 17–38. Lynne Rienner, Boulder.

Karmon, Yehuda
1966 A Geography of Settlement in Eastern Nigeria. Hebrew University Studies in Geography 15, Pamphlet 2. Jerusalem.

Kates, Robert W., Goran Hyden, and B. L. Turner II
1993 Theory, Evidence and Study Design. In *Population Growth and Agricultural Change in Africa*, edited by B. L. Turner II, G. Hyden, and R. W. Kates, pp. 1–40. University of Florida Press, Gainesville.

Katz, Yossi, and John C. Lehr
1991 Jewish and Mormon Agricultural Settlement in Western Canada: A Comparative Analysis. *Canadian Geographer* 35:128–142.

Keegan, William F., and Morgan D. Machlachlan
1989 The Evolution of Avunculocal Chiefdoms: A Reconstruction of Taino Kinship and Politics. *American Anthropologist* 91:613–630.

Kelley, Robert
1985 Hunter-Gatherer Mobility Strategies. *Journal of Anthropological Research* 39:277–306.

Killion, Thomas, editor
1992 *Gardens of Prehistory: The Archaeology of Settlement Agriculture in Greater Mesoamerica*. University of Alabama Press, Tuscaloosa.

Kimball, Solon T.
1950 Rural Social Organization and Co-operative Labor. *American Journal of Sociology* 55:38–49.

King, Leslie J.
1962 A Quantitative Expression of the Pattern of Urban Settlements in Selected Areas of the United States. *Tijdschrift Voor Con. en Soc. Geographie* 53:1–7.

1969 The Analysis of Spatial Form and Its Relation to Geographic Theory. *Annals of the Association of American Geographers* 59:573–595.

1984 *Central Place Theory*. Sage Publications, Beverly Hills, Calif.

King, Russell, and S. Burton

1982 Land Fragmentation: Notes on a Fundamental Rural Spatial Problem. *Progress in Human Geography* 6:475–494.

Kirch, Patrick V.

1978 Ethnoarchaeology and the Study of Agricultural Adaptation in the Humid Tropics. In *Explorations in Ethnoarchaeology*, edited by Richard A. Gould, pp. 103–125. University of New Mexico Press, Albuquerque.

1993 *The Wet and the Dry: Irrigation and Agricultural Intensification in Polynesia*. University of Chicago Press, Chicago.

Kjekshus, Helge

1977 The Tanzanian Villagization Policy: Implementation Lessons and Ecological Dimensions. *Canadian Journal of African Studies* 11: 269–282.

Kohler, Timothy A.

1992 Field Houses, Villages, and the Tragedy of the Commons in the Early Northern Anasazi Southwest. *American Antiquity* 57:617–635.

Kopytoff, Igor

1987 The Internal African Frontier: The Making of African Political Culture. In *The African Frontier: The Reproduction of Traditional African Societies*, edited by Igor Kopytoff, pp. 3–84. Indiana University Press, Bloomington.

Kowal, J. M., and A. W. Kassam

1978 *Agricultural Ecology of Savanna: A Study of West Africa*. Clarendon Press, Oxford.

Kristiansen, Kristian

1982 The Formation of Tribal Systems in Later European Prehistory: Northern Europe, 4000–500 B.C. In *Theory and Explanation in Archaeology*, edited by Colin Renfrew, M. J. Rowlands, and B. A. Segraves, pp. 241–280. Academic Press, New York.

Kroeber, A. L.

1917 The Superorganic. *American Anthropologist* 19:163–213.

Lagemann, Johannes

1977 *Traditional African Farming Systems in Eastern Nigeria: An Analysis of Reaction to Increasing Population Pressure*. Weltforum, Munich.

Layhe, Robert

1981 A Locational Model for Demographic and Settlement System Change: An Example from the American Southwest. Ph.D. diss., Department of Anthropology, Southern Illinois University, Carbondale.

Leach, Gerald

1976 *Energy and Food Production.* IPC Science and Technology Press, Guilford, Conn.

Lee, Ronald D.

1986 Malthus and Boserup: A Dynamic Synthesis. In *The State of Population Theory: Forward from Malthus,* edited by David Coleman and R. Schofield, pp. 96–130. Basil Blackwell, Oxford.

Lefferts, H. L., Jr.

1977 Frontier Demography: An Introduction. In *The Frontier: Comparative Studies,* vol. 1, edited by David H. Miller and J. O. Steffan, pp. 33–55. University of Oklahoma Press, Norman.

Lehr, John C.

1985 Kinship and Society in the Ukrainian Pioneer Settlement of the Canadian West. *Canadian Geographer* 29:207–219.

Lehr, John C., and Yossi Katz

1995 Crown, Corporation and Church: The Role of Institutions in the Stability of Pioneer Settlements in the Canadian West, 1870–1914. *Journal of Historical Geography* 21:413–427.

Lele, Uma, and Steven B. Stone

1989 *Population Pressure, the Environment, and Agricultural Intensification: Variations on the Boserup Hypothesis.* MADIA (Managing Agricultural Development in Africa) Paper 4. World Bank, Washington, D.C.

Lemon, James T.

1966 Agricultural Practices of National Groups in Eighteenth Century Southeastern Pennsylvania. *Geographical Review* 56:467–496.

Lewis, Kenneth E.

1984 *The American Frontier: An Archaeological Study of Settlement Pattern and Process.* Academic Press, New York.

Linares, Olga F.

1976 Garden Hunting in the American Tropics. *Human Ecology* 4:331–350.

1983 Social, Spatial and Temporal Relations: Diola Villages in Archaeological Perspective. In *Prehistoric Settlement Patterns: Essays in Honor of Gordon R. Willey,* edited by Evon Z. Vogt and Richard M. Leventhal, pp. 129–163. University of New Mexico Press, Albuquerque, and Harvard University, Cambridge.

1992 *Power, Prayer and Production.* Cambridge University Press, Cambridge.

Longacre, William A.

1970 *Archaeology as Anthropology: A Case Study.* Anthropological Papers of the University of Arizona 17. Tucson.

Lösch, August
1954 *The Economics of Location*. Yale University Press, New Haven. Originally published 1940.
LRDC (Land Resources Development Centre)
1981 *Land Resources of Central Nigeria, Agricultural Development Possibilities*, vol. 7, *An Atlas of Resource Maps*. Ministry of Overseas Development, Land Resources Development Centre, Land Resource Study 29. Surbiton, Surrey, U.K.
Mabogunje, Akin L.
1962 *Yoruba Towns*. University Press, Ibadan, Nigeria.
MacArthur, J. D.
1980 Some Characteristics of Farming in a Tropical Environment. In *Farming Systems in the Tropics*, edited by Hans Ruthenberg, pp. 19–29. Clarendon Press, Oxford.
MacLeod, W. N., D. C. Turner, and E. P. Wright
1971 The Geology of the Jos Plateau. In *Geological Survey of Nigeria Bulletin 32*, vol. 1, *General Geology*. Federal Republic of Nigeria, Lagos.
Madsen, Torsten
1982 Settlement Systems of Early Agricultural Societies in East Jutland, Denmark: A Regional Study of Change. *Journal of Anthropological Archaeology* 1:197–236.
Maos, Jacob O.
1984 *The Spatial Organization of New Land Settlement in Latin America*. Westview Press, Boulder.
Margolis, Maxine
1977 Historical Perspectives on Frontier Agriculture as an Adaptive Strategy. *American Ethnologist* 4:42–64.
Marshall, J. U.
1969 *The Location of Service Towns*. University of Toronto Press, Toronto.
Mayhew, Alan
1973 *Rural Settlement and Farming in Germany*. Barnes & Noble, New York.
McGovern, Thomas
1994 Management for Extinction in North Greenland. In *Historical Ecology: Cultural Knowledge and Changing Landscapes*, edited by Carole Crumley, pp. 127–154. School of American Research Press, Santa Fe.
McGuire, Randall H.
1984 The Boserup Model and Agricultural Intensification in the United States Southwest. In *Prehistoric Agricultural Strategies in the Southwest*, edited by Suzanne K. Fish and P. R. Fish, pp. 327–334. Arizona State University Anthropological Research Papers 33. Tempe.
McQuillan, D. Aidan
1978 Territory and Ethnic Identity: Some New Measures of an Old Theme

in the Cultural Geography of the United States. In *European Settlement and Development in North America,* edited by James R. Gibson, pp. 136–169. University of Toronto Press, Toronto.

Meek, C. K.

1931 *Tribal Studies in Northern Nigeria.* Oxford University Press, London.

Mikesell, Marvin W.

1960 Comparative Studies in Frontier History. *Annals of the Association of American Geographers* 50:62–74.

Moore, James A.

1983 The Trouble with Know-It-Alls: Information as a Social and Ecological Resource. In *Archaeological Hammers and Theories,* edited by J. Moore and A. Keene, pp. 173–191. Academic Press, New York.

Moore, M. P.

1975 Co-operative Labour in Peasant Agriculture. *Journal of Peasant Studies* 2:270–291.

Morgan, W. B.

1955a Farming Practice, Settlement Pattern and Population Density in South-Eastern Nigeria. *Geographical Journal* 121:320–333.

1955b The Change from Shifting to Fixed Settlement in Southern Nigeria. *Dept. of Geography, University of Ibadan, Research Notes (Papers)* 7:1–14.

1957 The "Grassland Towns" of the Eastern Region of Nigeria. *Transactions of the Institute of British Geographers* 23:213–224.

1969 The Zoning of Land Use around Rural Settlements in Tropical Africa. In *Environment and Land Use in Africa,* edited by M. F. Thomas and G. W. Whittington, pp. 301–319. Methuen, London.

1973 The Doctrine of the Rings. *Geography* 58:301–312.

Morgan, W.T.W.

1983 *Nigeria.* Longman, London.

Morrill, Richard L.

1962 Simulation of Central Place Patterns over Time. *Lund Studies in Geography, Series B (Human Geography)* 24:109–120.

1963a Development and Spatial Distribution of Towns in Sweden: An Historical-Predictive Approach. *Annals of the Association of American Geographers* 5:1–14.

1963b The Distribution of Migration Distances. *Regional Science Association Papers* 11:75–84.

Morrison, Kathleen D.

1994 The Intensification of Production: Archaeological Approaches. *Journal of Archaeological Method and Theory* 1:111–159.

1996 Typological Schemes and Agricultural Change: Beyond Boserup in Precolonial South India. *Current Anthropology* 37. In press.

Mortimore, Michael J.

1967 Land and Population Pressure in the Kano Close-Settled Zone,

Northern Nigeria. *Advancement of Science* 23:677–686. Reprinted in *People and Land in Africa South of the Sahara*, edited by R. M. Prothero, pp. 60–70. Oxford University Press, New York.

1968 Population Distribution, Settlement and Soils in Kano Province, Northern Nigeria, 1931–62. In *The Population of Tropical Africa*, edited by J. C. Caldwell and C. Okonjo, pp. 298–306. Columbia University Press, New York.

1993 The Intensification of Peri-Urban Agriculture: The Kano Close-Settled Zone, 1964–1986. In *Population Growth and Agricultural Change in Africa*, edited by B. L. Turner II, Goran Hyden, and Robert W. Kates, pp. 358–400. University of Florida Press, Gainesville.

Mulligan, Gordon F.

1981 Lösch's Single-Good Equilibrium. *Annals of the Association of American Geographers* 71:84–94.

1984 Agglomeration and Central Place Theory: A Review of the Literature. *International Regional Science Review* 9:1–42.

Murdock, George Peter

1967 *Ethnographic Atlas*. University of Pittsburgh Press, Pittsburgh.

Netting, Robert McC.

1965 Household Organization and Intensive Agriculture: The Kofyar Case. *Africa* 35:422–429.

1968 *Hill Farmers of Nigeria: Cultural Ecology of the Kofyar of the Jos Plateau*. University of Washington Press, Seattle.

1969 Ecosystems in Process: A Comparative Study of Change in Two West African Societies. *National Museum of Canada Bulletin* 230:102–112.

1973 Fighting, Forest and the Fly: Some Demographic Regulators among the Kofyar. *Journal of Anthropological Research* 29:164–179.

1974 Kofyar Armed Conflict: Social Causes and Consequences. *Journal of Anthropological Research* 30:139–163.

1987 Clashing Cultures, Clashing Symbols: Histories and Meanings of the Latok War. *Ethnohistory* 34:352–380.

1989 Smallholders, Householders, Freeholders: Why the Family Farm Works Well Worldwide. In *The Household Economy: Reconsidering the Domestic Mode of Production*, edited by Richard R. Wilk, pp. 221–244. Westview Press, Boulder.

1993 *Smallholders, Householders: Farm Families and the Ecology of Intensive, Sustainable Agriculture*. Stanford University Press, Stanford.

Netting, Robert McC., Glenn Davis Stone, and M. Priscilla Stone

1993 Agricultural Expansion, Intensification, and Market Participation among the Kofyar, Jos Plateau, Nigeria. In *Population Growth and Agricultural Change in Africa*, edited by B. L. Turner II, Goran Hyden, and Robert Kates, pp. 206–249. University of Florida Press, Gainesville.

Netting, Robert McC., M. Priscilla Stone, and Glenn D. Stone

1989 Kofyar Cash Cropping: Choice and Change in Indigenous Agricultural Development. *Human Ecology* 17:299–319.

Newman, Philip

1965 *Knowing the Gururumba*. Holt, Rinehart, & Winston, New York.

Neyman, J., and E. Scott

1957 On a Mathematical Theory of Populations Conceived as a Conglomeration of Clusters. *Cold Spring Harbor Symposia on Quantitative Biology* 22:109–120.

Nichols, Deborah L.

1987 Risk and Agricultural Intensification during the Formative Period in the Northern Basin of Mexico. *American Anthropologist* 89:596 616.

Norling, Gunnar

1960 Abandonment of Rural Settlement in Vasterbotten Lappmark, North Sweden, 1930–1960. *Geografiska Annaler* 42:232–243.

Norman, David L.

1969 Labour Inputs of Farmers: A Case Study of the Zaria Province of the North-Central State of Nigeria. *Nigerian Journal of Economic and Social Studies* 11:3–14.

Norman, D. W., E. B. Simmons, and H. M. Hays

1982 *Farming Systems in the Nigerian Savanna*. Westview, Boulder.

O'Brien, Michael J.

1984 *Grassland, Forest, and Historical Settlement*. University of Nebraska Press, Lincoln.

Offodile, Matthew E.

1976 *The Geology of the Middle Benue, Nigeria*. Publications of the Palaeontological Institution of the University of Uppsala, Special Volume 4. Uppsala, Sweden.

Ojo, G. J. Afolabi

1966 *Yoruba Culture: A Geographical Analysis*. University of London Press, London.

1973 Journey to Agricultural Work in Yorubaland. *Annals of the Association of American Geographers* 63:85–96.

Olsson, Gunnar

1966 Central Place Systems, Spatial Interaction, and Stochastic Processes. *Papers and Proceedings of the Regional Science Association* 18:13–45.

1968 Complementary Models: A Study of Colonization Maps. *Geografiska Annaler B* 50:115–132. Reprinted in *Readings in Rural Settlement Geography* (1975), edited by R. L. Singh and K. L. Singh, pp. 82–102. National Geographic Society of India, Varanasi.

Orcutt, Janet D., Eric Blinman, and Timothy A. Kohler

1990 Explanations of Population Aggregation in the Mesa Verde Region Prior to A.D. 900. In *Perspectives on Southwestern Prehistory*, edited by

Paul E. Minnis and Charles L. Redman, pp. 196–212. Westview Press, Boulder.

Ostergren, Robert

1979 A Community Transplanted: The Formative Experience of a Swedish Immigrant Community in the Upper Middle West. *Journal of Historical Geography* 5:189–212.

Padoch, Christine

1985 Labor Efficiency and Intensity of Land Use in Rice Production: An Example from Kalimantan. *Human Ecology* 13:271–289.

1986 Agricultural Site Selection among Permanent Field Farmers: An Example from East Kalimantan, Indonesia. *Journal of Ethnobiology* 6:279–288.

Parsons, Jeffrey R.

1972 Archaeological Settlement Patterns. *Annual Review of Anthropology* 1:127–150.

Paynter, Robert

1982 *Models of Spatial Inequality: Settlement Patterns in Historical Archaeology*. Academic Press, New York.

Perham, Margery

1968 *Lugard: The Years of Authority, 1989–1945*. Archdon Books, Hamden, Conn.

Pimentel, David, and Marcia Pimentel

1979 *Food, Energy and Society*. John Wiley and Sons, New York.

Pingali, Prabhu, Yves Bigot, and Hans P. Binswanger

1987 *Agricultural Mechanization and the Evolution of Farming Systems in Sub-Saharan Africa*. Johns Hopkins University Press, for the World Bank, Baltimore.

Plog, Fred T., and James Hill

1971 Explaining Variability in the Distribution of Sites. In *The Distribution of Prehistoric Population Aggregates*, edited by George J. Gumerman, pp. 7–36. Prescott College Anthropological Reports 1. Prescott, Ariz.

Preston, Richard E.

1985 Christaller's Neglected Contribution to the Study of the Evolution of Central Places. *Progress in Human Geography* 1:181–193.

Preucel, Robert W.

1987 Settlement Succession on the Pajarito Plateau, New Mexico. *Kiva* 53:3–33.

Price, Barbara

1977 Shifts in Production and Organization: A Cluster-Interaction Model. *Current Anthropology* 18:209–233.

Price, Cynthia R., and James E. Price

1981 Investigation of Settlement and Subsistence Systems in the Ozark Border Region of Southeast Missouri during the First Half of the 19th Century: The Widow Harris Cabin Project. *Ethnohistory* 28:237–258.

PRO (Public Records Office)

1918 File 44654, Concerning Incident near Bauchi-Muri Border. Public Records Office, Kew Gardens, England.

1939 Letter Dated 22 Aug 1939 from the Governor's Deputy to the Secretary of State for the Colonies, Concerning Return of Hill Dimmuks. CO S83/244/30442 SS193, Public Records Office, Kew Gardens, England.

Protectorate of Northern Nigeria

1901 *Annual Report for 1900*. Colonial Office, London.

Prothero, R. Mansell

1957 Land Use at Soba, Zaria Province, Northern Nigeria. *Economic Geography* 33:72–86.

Pryor, Frederic L.

1985 The Invention of the Plow. *Comparative Studies in Society and History* 27:727–743.

1986 The Adoption of Agriculture: Some Theoretical and Empirical Evidence. *American Anthropologist* 88:879–897.

Pryor, Frederic L., and Stephen B. Maurer

1982 On Induced Economic Change in Precapitalist Economies. *Journal of Development Economics* 10:325–353.

Puleston, Dennis E., and Olga S. Puleston

1971 An Ecological Approach to the Origins of Maya Civilization. *Archaeology* 24(0):330–337.

Rappaport, Roy A.

1968 *Pigs for the Ancestors*. Yale University Press, New Haven.

Reid, J. Jefferson

1985 Measuring Social Complexity in the American Southwest. In *Status, Structure and Organization: Current Archaeological Approaches*, edited by M. Thompson, M. T. Garcia, and F. J. Kense, pp. 168–172. Proceedings of the 1983 Chacmool Conference, Calgary.

Renfrew, Colin

1972 *The Emergence of Civilisation*. Methuen, London.

Reynolds, H. T.

1977 *The Analysis of Cross-Classifications*. Free Press, Riverside, N.J.

Rice, J. G.

1977 The Role of Culture and Community in Frontier Prairie Farming. *Journal of Historical Geography* 3:155–175.

Richards, Paul

1978a Farming Systems, Settlement and State Formation: The Nigerian Evidence. In *Social Organization and Settlement: Contributions from Anthropology, Archaeology and Geography*, edited by David Green, C.

Haselgrove, and M. Spriggs, pp. 477–509. BAR International Series 47. London.

1978b Spatial Organisation and Social Change in West Africa—Notes for Historians and Anthropologists. In *The Spatial Organisation of Culture*, edited by Ian Hodder, pp. 271–289. University of Pittsburgh Press, Pittsburgh.

1983 Farming Systems and Agrarian Change in West Africa. *Progress in Human Geography* 7:1–39.

1985 *Indigenous Agricultural Revolution: Ecology and Food Production in West Africa*. Hutchinson, London.

Richardson, Bonham C.

1974 Distance Regularities in Guayanese Rice Cultivation. *Journal of Developing Areas* 8:235–256.

Riley, Thomas J., and Glen Freimuth

1979 Field Systems and Frost Drainage in the Prehistoric Agriculture of the Upper Great Lakes. *American Antiquity* 1979:271–285.

Roberts, Brian K.

1977 *Rural Settlement in Britain*. Dawson/Archon, Kent.

Robertson, A. F.

1987 *The Dynamics of Productive Relationships: African Share Contracts in Comparative Perspective*. Cambridge University Press, Cambridge.

Robinson, W., and A. Schutjer

1984 Agricultural Development and Cultural Change: A Generalization of the Boserup Model. *Economic Development and Cultural Change* 32:355–366.

Roet, Jeffrey B.

1985 Land Quality and Land Alienation on the Dry Farming Frontier. *Professional Geographer* 37:173–183.

Roper, Donna C.

1979 The Method and Theory of Site Catchment Analysis: A Review. In *Advances in Archaeological Method and Theory*, vol. 2, edited by Michael B. Schiffer, pp. 119–140. Academic Press, New York.

Rossman, David L.

1976 A Site Catchment Analysis of San Lorenzo, Veracruz. In *The Early Mesoamerican Village*, edited by Kent V. Flannery, pp. 95–103. Academic Press, New York.

Rowlands, M. J.

1972 Defence: A Factor in the Organization of Settlements. In *Man, Settlement and Urbanism*, edited by Peter J. Ucko, R. Tringham, and G. W. Dimbleby, pp. 447–462. Schenkman, Cambridge, Mass.

Rowling, C. W.

1946 Report on Land Tenure in Plateau Province. National Archives Kaduna, File 997/S.1.

Rubin, Julius
1973 Notes on the Comparative Study of the Agriculture of World Regions. *Peasant Studies Newsletter* 2(0):1–4.
Ruthenberg, Hans
1980 *Farming Systems in the Tropics*. 3d ed. Clarendon Press, Oxford.
Rutter, Andrew F.
1971 Ashanti Vernacular Architecture. In *Shelter in Africa*, edited by Paul Oliver, pp. 153–171. Barrie & Jenkins, London.
Sahlins, Marshall D.
1961 The Segmentary Lineage: An Organization of Predatory Expansion. *American Anthropologist* 63:322–345.
1972 *Stone Age Economics*. Aldine-Atherton, Chicago.
Salamon, Sonya
1985 Ethnic Communities and the Structure of Agriculture. *Rural Sociology* 50:323–340.
Sallade, Jane K., and David P. Braun
1982 Spatial Organization of Peasant Agricultural Subsistence Territories: Distance Factors and Crop Location. In *Ethnography by Archaeologists*, edited by Elisabeth Tooker, pp. 19–41. American Ethnological Society, Washington, D.C.
Salmon, Merrilee H.
1982 *Philosophy and Archaeology*. Academic Press, New York.
Salmon, Merrilee H., and Wesley C. Salmon
1979 Alternative Models of Scientific Explanation. *American Anthropologist* 81:61–74.
Sanders, William T.
1967 Settlement Patterns. *Handbook of Middle American Indians* 6:53–86.
1973 The Cultural Ecology of the Lowland Maya: A Reevaluation. In *The Classic Maya Collapse*, edited by T. Patrick Culbert, pp. 325–365. University of New Mexico Press, Albuquerque.
1981 Classic Maya Settlement Patterns and Ethnographic Analogy. In *Lowland Maya Settlement Patterns*, edited by Wendy Ashmore, pp. 351–369. University of New Mexico Press, Albuquerque.
Sanders, William T., and Thomas W. Killion
1992 Factors Affecting Settlement Agriculture in the Ethnographic and Historic Record of Mesoamerica. In *Gardens of Prehistory*, edited by Thomas W. Killion, pp. 14–31. University of Alabama Press, Tuscaloosa.
Sanders, William T., and Deborah L. Nichols
1988 Ecological Theory and Cultural Evolution in the Valley of Oaxaca. *Current Anthropology* 29:33–80.

Sanders, William T., Jeffrey Parsons, and Robert S. Santley
1979 *The Basin of Mexico: Ecological Processes in the Evolution of a Civilization.* Academic Press, New York.

Sanders, William T., and David Webster
1978 Unilinealism, Multilinealism, and the Evolution of Complex Societies. In *Social Archaeology: Beyond Subsistence and Dating,* edited by Charles L. Redman, M. J. Berman, et al., pp. 249–302. Academic Press, New York.

Saul, Mahir
1983 Work Parties, Wages and Accumulation in a Voltaic Village. *American Ethnologist* 10:77–96.

Savage, William W., and Stephen I. Thompson
1979 The Comparative Study of the Frontier: An Introduction. In *The Frontier: Comparative Studies,* vol. 2, edited by W. W. Savage and S. I. Thompson, pp. 3–24. University of Oklahoma Press, Norman.

Schiffer, Michael B.
1976 *Behavioral Archaeology.* Academic Press, New York.

Schiffer, Michael B., and Randall H. McGuire
1982 The Study of Cultural Adaptations. In *Hohokam and Patayan: Prehistory of Southwestern Arizona,* edited by Randall H. McGuire and Michael B. Schiffer, pp. 223–274. Academic Press, New York.

Schluter, M. G., and T. D. Mount
1976 Some Management Objectives for the Peasant Farmer: An Analysis of Risk Aversion in the Choice of Cropping Patterns. *Journal of Development Studies* 12:246–267.

Schultz, John F.
1976 Population and Agricultural Change in Nigerian Hausaland. Ph.D. diss., Department of Geography, Columbia University, New York.

Schultz, T. W.
1964 *Transforming Traditional Agriculture.* Yale University Press, New Haven.

Semple, Ellen C.
1932 *The Geography of the Mediterranean Region: Its Relation to Ancient History.* Constable, London.

Shankman, Paul
1991 Culture Contact, Cultural Ecology, and Dani Warfare. *Man* 26:299–322.

Shyrock, R. H.
1939 British versus German Traditions in Colonial Agriculture. *Mississippi Valley Historical Review* 26:39–54.

Siddle, David J.
1970 Location Theory and the Subsistence Economy: The Spacing of

Rural Settlements in Sierra Leone. *Journal of Tropical Geography* 31:79–90.

Silberfein, Marilyn

1972 A Cyclic Approach to Settlement Patterns in Africa. *S.A. Journal of African Affairs* 3(0):11–30.

1989 Settlement Form and Rural Development: Scattered versus Clustered Settlement. *Tijdshrift Voor Economishe en Soc. Geografie* 80:258–268.

Sithole, Masirupa

1985 The Salience of Ethnicity in African Politics: The Case of Zimbabwe. *Journal of Asian and African Studies* 20:3–4.

Skinner, G. William

1964 Marketing and Social Structure in Rural China. *Journal of Asian Studies* 24:3–43.

1977 Cities and the Hierarchy of Local Systems. In *The City in Late Imperial China*, edited by G. William Skinner, pp. 275–352. Stanford University Press, Stanford.

Smith, Carol A.

1975 Production in Western Guatemala: A Test of von Thünen and Boserup. In *Formal Methods in Economic Anthropology*, edited by Stuart Plattner, pp. 5–37. American Anthropological Association Publication 4. Washington, D.C.

Smith, E. G.

1975 Fragmented Farms in the United States. *Annals of the Association of American Geographers* 65:58–70.

Smith, Michael E.

1979 The Aztec Marketing System and Settlement Pattern in the Valley of Mexico: A Central Place Analysis. *American Antiquity* 44:110–125.

1980 The Role of the Marketing System in Aztec Society and Economy: Reply to Evans. *American Antiquity* 45:876–883.

Smith, Philip E. L.

1972 Land-Use, Settlement Patterns and Subsistence Agriculture: A Demographic Perspective. In *Man, Settlement and Urbanism*, edited by Peter J. Ucko, R. Tringham, and G. W. Dimbleby, pp. 409–425. Schenkman, Cambridge, Mass.

Smith, Philip E. L., and T. Cuyler Young, Jr.

1972 The Evolution of Early Agriculture and Culture in Greater Mesopotamia: A Trial Model. In *Population Growth: Anthropological Implications*, edited by Brian Spooner, pp. 1–59. MIT Press, Cambridge.

Smith, Roger W.

1974 Sketches of a Dynamic Central Place Theory. *Economic Geography* 50:219–227.

1977 Dynamic Central Place Theory: Results of a Simulation Approach. *Geographical Analysis* 9:226–243.

Stark, Barbara L., and Dennis L. Young

1981 Linear Nearest Neighbor Analysis. *American Antiquity* 46:284–300.

Steponaitis, Vincas P.

1981 Settlement Hierarchies and Political Complexity in Nonmarket Societies: The Formative Period of the Valley of Mexico. *American Anthropologist* 83:320–363.

Steward, Julian H.

1955 *Theory of Culture Change*. University of Illinois Press, Urbana.

Stone, Glenn Davis

1983 Material Correlates of the Developmental Cycle of the Household: An Ethnoarchaeological Case Study. Paper at the meeting of the Society for American Archaeology, Pittsburgh.

1988 Agrarian Ecology and Settlement Patterns: An Ethnoarchaeological Case Study. Ph.D. diss., Department of Anthropology, University of Arizona, Tucson.

1991a Agricultural Territories in a Dispersed Settlement System. *Current Anthropology* 32:343–353.

1991b Settlement Ethnoarchaeology: Changing Patterns among the Kofyar of Nigeria. *Expedition* 33(1):16–23.

1992 Social Distance, Spatial Relations, and Agricultural Production among the Kofyar of Namu District, Plateau State, Nigeria. *Journal of Anthropological Archaeology* 11:152–172.

1993a Agricultural Abandonment: A Comparative Study in Historical Ecology. In *The Abandonment of Settlements and Regions: Ethnoarchaeological and Archaeological Approaches*, edited by C. Cameron and S. Tomka, pp. 74–81. Cambridge University Press, New York.

1993b Agrarian Settlement and the Spatial Disposition of Labor. In *Spatial Boundaries and Social Dynamics: Case Studies from Food-Producing Societies*, edited by A. Holl and T. Levy, pp. 25–38. International Monographs in Prehistory, Ann Arbor.

1994 Agricultural Intensification and Perimetric Features: Ethnoarchaeological Evidence from Nigeria. *Current Anthropology* 35:317–324.

1996 Settlement Concentration and Dispersal among the Kofyar. In *Rural Settlement Structure and African Development*, edited by M. Silberfein. Westview Press, Boulder, in press.

Stone, Glenn Davis, M. P. Johnson-Stone, and R. M. Netting

1984 Household Variability and Inequality in Kofyar Subsistence and Cash-Cropping Economies. *Journal of Anthropological Research* 40: 90–108.

Stone, Glenn Davis, Robert McC. Netting, and M. Priscilla Stone
1990 Seasonality, Labor Scheduling and Agricultural Intensification in the Nigerian Savanna. *American Anthropologist* 92:7–24.

Stone, M. Priscilla
1988a Women Doing Well: A Restudy of the Nigerian Kofyar. *Research in Economic Anthropology* 10:287–306.
1988b Women, Work and Marriage: A Restudy of the Nigerian Kofyar. Ph.D. diss., Department of Anthropology, University of Arizona, Tucson.

Stone, M. Priscilla, Glenn Davis Stone, and Robert McC. Netting
1995 The Sexual Division of Labor in Kofyar Agriculture. *American Ethnologist* 22:165–186.

Struever, Stuart
1968 Woodland Subsistence-Settlement Systems in the Lower Illinois Valley. In *New Perspectives in Archeology*, edited by Sally R. Binford and L. R. Binford, pp. 285–312. Aldine, New York.

Suttles, Gerald D.
1972 *The Social Construction of Communities*. University of Chicago, Chicago.

Swedlund, Alan C.
1975 Population Growth and Settlement Pattern in Franklin and Hampshire Counties, Massachusetts, 1650–1850. In *Population Studies in Archaeology and Biological Anthropology: A Symposium*, edited by Alan C. Swedlund, pp. 22–33. Society for American Archaeology Memoir 30. Issued as *American Antiquity* 40, no. 2.

Temple, C. L., editor
1922 *Notes on the Tribes, Provinces, Emirates and States of the Northern Provinces of Nigeria*. 2d ed. Barnes & Noble, New York.

Thomas, David Hurst
1972 Western Shoshone Ecology: Settlement Patterns and Beyond. In *Great Basin Cultural Ecology: A Symposium*, edited by D. D. Fowler, pp. 135–153. Desert Research Institution Publications in Social Science no. 8.
1988 Diversity in Hunter-Gatherer Cultural Geography. In *Diversity in Archaeology*, edited by Robert D. Leonard and G. T. Jones, pp. 87–94. Cambridge University Press, Cambridge.

Thomas, R. W., and R. J. Huggett
1980 *Modelling in Geography: A Mathematical Approach*. Barnes & Noble, Totowa, N.J.

Trigger, Bruce
1968 The Determinants of Settlement Patterns. In *Settlement Archaeology*, edited by K. C. Chang, pp. 53–78. National Press, Palo Alto, Calif.

Tripp, Robert B.
1982 Time Allocation in Northern Ghana: An Example of the Random Visit Method. *Journal of Developing Areas* 16:391–400.

Troll, C.
1966 *Seasonal Climates of the Earth*. Springer, Berlin.

Turner, B. L., and W. E. Doolittle
1978 The Concept and Measure of Agricultural Intensity. *Professional Geographer* 30:297–301.

Turner, B. L., R. Q. Hanham, and A. V. Portararo
1977 Population Pressure and Agricultural Intensity. *Annals of the Association of American Geographers* 67:384–396.

Turner, B. L., II, and S. B. Brush, editors
1987 *Comparative Farming Systems*. Guilford Press, New York.

Turner, B. L., II, G. Hyden, and R. Kates, editors
1993 *Population Growth and Agricultural Intensification in Africa*. University of Florida Press, Gainesville.

Turner, Frederick Jackson
1920 *The Frontier in American History*. Henry Holt, New York.

Udo, Reuben K.
1961 Land and Population in Otoro District. *Nigerian Geographical Journal* 4:3–19.
1963 Patterns of Population Distribution and Settlement in Eastern Nigeria. *Nigerian Geographical Journal* 6:73–88.
1965 Disintegration of Nucleated Settlement in Eastern Nigeria. *Geographical Review* 55:53–67.
1966 Transformation of Settlement in British Tropical Africa. *Nigerian Geographical Journal* 9:129–144.
1970 *Geographical Regions of Nigeria*. Heinemann, London.

Unomah, A. Chukwudi
1982 The Lowlands Salt Industry. In *Studies in the History of Plateau State, Nigeria*, edited by Elizabeth Isichei, pp. 151–178. Macmillan, London.

Upham, Steadman
1985 Interpretations of Prehistoric Political Complexity in the Central and Northern Southwest. In *Status, Structure and Organization: Current Archaeological Approaches*, edited by M. Thompson, M. T. Garcia, and F. J. Kense, pp. 175–180. Proceedings of the 1983 Chacmool Conference, Calgary.

Vasey, Daniel E.
1979 Population and Agricultural Intensity in the Humid Tropics. *Human Ecology* 7:269–283.

Vermeer, Donald E.

1970 Population Pressure and Crop Rotational Changes among the Tiv of Nigeria. *Annals of the Association of American Geographers* 60:299–314.

Vickers, William T.

1988 Game Depletion Hypothesis of Amazonian Adaptation: Data from a Native Community. *Science* 239:1521–1522.

Vita-Finzi, C., and E. S. Higgs

1970 Prehistoric Economy in the Mount Carmel Area of Palestine: Site Catchment Analysis. *Proceedings of the Prehistoric Society* 36:1–37.

Vivian, R. Gwinn

1989 Kluckhohn Reappraised: The Chacoan System as an Egalitarian Enterprise. *Journal of Anthropological Research* 45:101–113.

Vogt, Evon Z.

1956 An Appraisal of "Prehistoric Settlement Patterns in the New World." In *Prehistoric Settlement Patterns in the New World*, edited by Gordon R. Willey, pp. 173–182. Viking Fund Publications 23. Wenner-Gren Foundation, New York.

1983 Ancient and Contemporary Maya Settlement Patterns: A New Look from the Chiapas Highlands. In *Prehistoric Settlement Patterns: Essays in Honor of Gordon R. Willey*, edited by Evon Z. Vogt and Richard M. Leventhal, pp. 89–114. University of New Mexico Press, Albuquerque.

von Thünen, Johann Heinrich

1966 *Von Thünen's Isolated State*. Translated by Carla M. Wartenburg, edited by P. Hall. Pergamon Press, London. Originally published 1826.

Voorhies, Barbara

1982 An Ecological Model of the Early Maya of the Central Lowlands. In *Maya Subsistence*, edited by Kent V. Flannery, pp. 65–95. Academic Press, New York.

Waddell, E.

1972 *The Mound Builders: Agricultural Practices, Environment, and Society in the Central Highlands of New Guinea*. University of Washington Press, Seattle.

Warren, Robert E., and Michael J. O'Brien

1984 A Model of Frontier Settlement. In *Grassland, Forest and Historical Settlement*, edited by Michael J. O'Brien, pp. 22–57. University of Nebraska Press, Lincoln.

Watson, Patty Jo, Steven A. LeBlanc, and Charles L. Redman

1971 *Archaeological Explanation: The Scientific Method in Archaeology*. Columbia University Press, New York.

Whallon, Robert, Jr.

1968 Investigations of Late Prehistoric Social Organization in New York

State. In *New Perspectives in Archeology*, edited by Sally R. Binford and L. R. Binford, pp. 223–244. Aldine, New York.

Wharton, C. R., Jr., editor

1969 *Subsistence Agriculture and Economic Development.* Aldine, Chicago.

White, Michael J.

1983 The Measurement of Spatial Segregation. *American Journal of Sociology* 88:1008–1018.

White, Roger W.

1977 Dynamic Central Place Theory: Results of a Simulation Approach. *Geographical Analysis* 10:201–208.

Wilhelm, Eugene J.

1978 Folk Settlements in the Blue Ridge Mountains. *Appalachian Journal* 5(2):204–245.

Wilk, Richard R.

1985 Dry Season Agriculture among the Kekchi Maya and Its Implications for Prehistory. In *Prehistoric Lowland Maya Environment and Subsistence Economy*, edited by Mary Pohl, pp. 47–57. Peabody Museum Papers 77. Cambridge, Mass.

1988 Ancient Maya Household Organization: Evidence and Analogies. In *Household and Community in the Mesoamerican Past*, edited by Richard Wilk and Wendy Ashmore, pp. 135–152. University of New Mexico Press, Albuquerque.

1991 *Household Ecology: Economic Change and Domestic Life among the Kekchi Maya in Belize*. University of Arizona Press, Tucson.

Wilk, Richard R., and Robert McC. Netting

1984 Households: Changing Forms and Functions. In *Households: Comparative and Historical Studies of the Domestic Group*, edited by Robert McC. Netting, R. R. Wilk, and E. J. Arnould, pp. 1–28. University of California Press, Berkeley.

Wilkinson, J. C.

1977 *Water and Tribal Settlement in South-East Arabia.* Clarendon Press, Oxford.

Willey, Gordon R.

1953 *Prehistoric Settlement Patterns in the Virú Valley, Perú.* Smithsonian Institute Bureau of American Ethnology Bulletin 15. Washington, D.C.

1974 The Virú Valley Settlement Pattern Study. In *Archaeological Researches in Retrospect*, edited by Gordon R. Willey, pp. 149–176. Winthrop, Cambridge, Mass.

Williams, C. N., and K. T. Joseph

1973 *Climate, Soil and Crop Production in the Humid Tropics.* Oxford University Press, Singapore.

Williams, Glyn

1977 Differential Risk Strategies as Cultural Style among Farmers in the Lower Chubut Valley, Patagonia. *American Ethnologist* 4:65–83.

Wilshusen, Richard, and Glenn Davis Stone

1990 An Ethnoarchaeological Perspective on Soils. *World Archaeology* 22:104–114.

Winters, Howard D.

1968 *The Riverton Culture*. Illinois State Museum, Springfield.

Wolf, Eric R.

1966 *Peasants*. Prentice-Hall, Englewood Cliffs, N.J.

1994 Perilous Ideas: Race, Culture, People. *Current Anthropology* 35:1–12.

Wood, J.

1971 Fitting Discrete Probability Distributions to Prehistoric Settlement Patterns. In *The Distribution of Prehistoric Population Aggregates*, edited by George J. Gumerman, pp. 63–82. Prescott College Anthropological Reports 1. Prescott, Ariz.

Zarky, Alan

1976 Statistical Analysis of Site Catchments at Ocos, Guatemala. In *The Early Mesoamerican Village*, edited by Kent V. Flannery, pp. 117–130. Academic Press, New York.

Zent, Stanford R.

1992 Historical and Ethnographic Ecology of the Upper Cuao River Wothiha: Clues for an Interpretation of Native Guianese Social Organization. Ph.D. diss., Department of Anthropology, Columbia University, New York.

Index

abandonment, 7, 21, 25, 34, 40, 51, 150, 165, 184, 187, 194, 207: of early bush farms, 85; of Kofyar homeland, 79; profiles, 166–174, 211; in tropical vs. temperate climates, 44, 45
adaptive strategy, 186, 191
Africa, settlement patterns: Ghana, 56; Hausa, 3, 46, 47, 59, 97, 119; Ibo, 16, 47, 163, 176, 179; Ngwa, 16, 163; Nigeria, 45, 175; Nigerian downhill movements, 96; Senegal, 64; Tanzania, 45; Yoruba, 50
agrarian settlement: in Central Place Theory, 4, 5; defined, 5
agricultural collaboration, 3, 9, 44, 53, 95, 114, 136: cited as cause of nucleation, 48; effects on settlement pattern, 55, 114–117, 120–123, 128, 142, 143, 183; and ethnicity, 56, 142, 143; festive vs. exchange labor defined, 54, 109, 209; neglected by analysts, 21, 54; tasks favoring, 54, 55, 109–112, 144; in *toenglu*, 4, 66. *See also mar muos; wuk*
agricultural intensification, 4, 28, 43: defined, 29; on frontier, 102, 104–115, 121, 194; hydraulic corollary, 163; intensification slopes, 30–41, 45, 86, 165, 166, 172, 175, 185, 202; labor efficiency, 32, 115, 193, 202; and risk avoidance, 34; sexual division of labor, 108, 110, 164; types of, 30–32. *See also* Boserup, E.
agricultural movement, 42, 129–133: depiction of, 134; and distance thresholds, 131–136, 138; effects in extensive farming, 44, 126; effects on dispersal, 43, 121, 125, 126; effects on farm shape, 55, 120, 122; and ethnicity, 138, 139
Amazon, settlement patterns, 21, 39, 47, 48, 51, 130, 203, 213
Andes, settlement patterns, 5, 51, 128
animals: effects on settlement, 91, 95, 146, 147, 188, 207
Ankwe. *See* Goemai
archaeology: European Neolithic settlement, 32, 45; Formative Mesoamerican settlement, 6, 7, 21; geographical theory in, 10, 23, 24, 26; LBK culture, 21; models of settlement systems, 6, 12, 23; New

Archaeologists, 6; settlement theory, 5, 9, 10, 19. *See also* Binford, L.; Flannery, K.; territorial analysis
architecture, domestic, 3, 4, 64, 65, 92–95, 119, 127, 188, 199, 207, 212
Asia, Southeast, settlement patterns, 32, 33, 48–51, 91, 175, 207, 211

Bauchi dynasty, 77
beer, millet *(muos)*, 62, 63, 104, 109, 110, 114, 115, 184, 202–206, 209: forbidden by Protestants, 113, 126, 127; sale of, 104, 176, 177
Benue Piedmont, 74–76, 165–168
Benue Valley: agricultural migration to, 3, 80; slave raiding in, 59, 77, 206
Binford, L., collector-forager model, 5, 6, 51
Boserup, E., 10, 28–30, 203: empirical tests of model, 31, 33, 46–49, 202; model of agricultural change, 10, 28–32, 34, 35; problem of environmental variation, 31, 35–39, 175
Brookfield, H., 29, 30, 34
burials, 4, 84

catchment analysis. *See* territorial analysis
Central Place Theory, 4, 17–19, 50, 195, 209
China, settlement patterns, 47
Chisholm, M., 15, 43, 159: model of cultivation radius, 13–17, 129, 130, 132, 163, 204. *See also* von Thünen, H.
chitemene agriculture, 36, 38
colonial government: archives in Nigeria, 77; effect on infrastructure, 80, 81, 206; effect on Kofyar settlement, 79, 82; effect on Tiv settlement, 188; ethnic/political creations of, 63, 66–68, 72, 73, 99, 206; initial encounter with Kofyar, 60; *reducciones*, 51; taxation, 81, 207; urban growth caused by, 81
commercial vs subsistence economies, 15, 19, 34, 80, 81, 202
core area, 92, 97, 101–104, 114, 119, 128, 146: defined, 91
cultural anthropology: settlement studies in, 10

defensive settlement, 47, 50, 82, 128: on frontier, 94, 146; in Jos Plateau, 59; slaving, 57
determinants of settlement: agricultural collaboration, 120–125, 128; in Chisholm's "inverted analysis," 15; in commercial vs. subsistence economies, 15, 19, 41; effects of prior settlements, 21, 22, 24, 96, 182; probabilistic models, 8; social factors, 7, 125–127
disease: effects on settlement, 82
dispersal: with agricultural intensification, 43, 46–50, 121, 128, 177, 204; defined, 43; and inefficiency, 44; political factors, 48; political factors militating against, 50, 51; produced by farm consolidation, 23; to protect land, 43, 51, 91, 95; and social relations, 9, 24, 43, 44, 48
Doedel (Chief of Kwa), 85–87, 97, 99, 127, 146, 207
Doemak, 60: character of, 206; chief of, 68, 86; construed as a tribe, 60, 68; early bush farming, 84–86, 99, 144; geographical area, 60, 67, 80; Hill Doemaks, 82; market, 80; *sargwat*, 71, 144, 146, 152, 205; supravillage, 66; tax rate, 81

Doka, 60, 66–69, 72, 73, 144, 205
Doolittle, W., 29, 53, 202

ecology: of crops, 104, 105, 113, 114; of fallows, 38, 39; in settlement models, 22–25, 192
equifinality, 7
ethnic gravity. *See* gravity models
ethnicity, 56, 146: affected by colonialism, 67, 68, 73; on American frontier, 96, 140–143, 147; and geography, 64, 68, 71, 205, 210; and marriage patterns, 69, 71, 72; and site encystment, 152; as social propinquity, 63, 67–73, 138–141, 147, 156; in ungwas, 97, 145, 146, 151; variation among American Indians, 63. *See also* sociosettlement units; tribe
Europe, settlement patterns, 24, 26, 29, 44, 45, 48, 54, 175, 190–192, 203, 204: "agro-towns," 49; Finland, 190; Germany, 4, 5, 18; Greece, 16; LBK culture, 21, 45, 211; Sweden, 23
extensive agriculture. *See* shifting cultivation

farm fragmentation, 43, 49, 50, 115, 209
festive labor. *See* agricultural collaboration; *mar muos*
Flannery, K., 6–8, 21, 120, 138, 181, 194, 201
Fried, M., 68
frontier: agricultural intensification, 101, 104, 122; American, 96, 140; as a context for theory building, 8–11, 82, 88, 122; early agriculture, 89, 90, 94, 95, 146; households, 94, 95, 102, 120
Fulani, 3, 59, 77, 79, 162

Ganguk *sargwat*, 67, 68, 71, 145, 146, 152, 174, 205: translation, 206
Gankogom *sargwat*, 67, 72, 73, 144–147, 151, 152, 205, 210: translation, 205
geography, 26: Hudson model of settlement evolution, 10; settlement studies, 10, 12, 18, 26, 27. *See also* Brookfield, H.; Central Place Theory; Doolittle, W.; Hudson, J.; Jordan, T.; Turner, B.; von Thünen, H.
Goejak, 99, 114, 136, 151
Goemai (Ankwe), 77, 83, 206
Goewan gari, 92–97, 127, 146, 161
gravity models, 20, 22, 44, 183: ethnic gravity, 141–144

households: assumed to be independent, 54; in dispersed settlement systems, 43; effects of production regime on, 24, 62, 95; fragmented, 92, 94, 102; income, 176; Kofyar, developmental cycle, 65, 124, 199, 211; Kofyar, head *(wupinlu)*, 93, 100, 105, 109, 164, 165, 199, 211; Kofyar, size, 64, 65, 86, 87, 99, 122, 125, 194; Kofyar, size vs. compound size, 64, 65, 121, 199; as labor pools, 53, 54, 62, 65, 87, 95, 109–111, 142; relationship to farm size, 25, 205; Tiv, size, 95, 96
Hudson, J., 10, 22–27, 120, 201
hunter-gatherers, 5, 16, 51

Iban, 51, 207, 211
intensive agriculture. *See* agricultural intensification
Islam: Fulani jihad, 59; Kofyar conversion to, 86, 99, 119, 127, 210; Muslim emirates, 77

Jangwa Clay Plains, 75, 76, 160, 166, 171
Jipal *sargwat*, 67, 68, 71, 72, 206
Jordan, T., 24, 190–192
Jukun empire, 77, 206

Kofyar: compared to Tiv, 84, 97, 98, 187, 188, 212; language, 188; origin myth, 60, 69, 77, 205; subdued by British, 60
Kofyar homeland: agriculture, 60–62, 80; effect on frontier settlement, 3, 4, 8, 96; ethnicity in, 64, 68, 69–72, 205; households, 64, 65, 87; map, 61; markets, 80; population density, 60, 207; reasons for abandonment, 79, 80, 82, 87; *sargwat* (warfare alliances), 67; settlement pattern in, 3, 4, 60–67; *toenglu*, 65, 66, 205; villages, 66; warfare, 60, 67, 80, 206
kop, defined, 97
Kwalla, 60, 66–68, 71, 80–84, 144, 197, 205
Kwallala, 107, 114, 115, 130–136, 177, 178, 199, 210

labor demands of production, 62, 95: effects on settlement, 93, 95; grains, 104, 105, 112, 113, 203; linear vs. simultaneous, 54, 111, 112, 144; overall inputs, 107, 208; quality of labor, 111–113; rice, 105; in savanna, 90, 105, 108; yams, 105, 203
Lafia, 74, 76, 85: emirate, 77; raiders from, 59, 77
land tenure, 4, 43, 48–51, 189
Langkaku, 84, 85, 147, 160–163, 171, 213
Linares, O., 64
locational intensification, 52–53, 164–165
locational optimization: alternatives, 186, 191–195; farm shape, 122, 123; land use, 13–15; site spacing, 123
locational optimization, factors overriding: attraction to towns, 48–51; defensive considerations, 50, 51; "economy of affection," 141, 142; religion, 119, 125; social organization of labor, 134–137, 142; symbolic aspects of settlements, 56; unspecified "social" factors, 6; varying in strength, 8, 9, 56

macrocompounds: on frontier, 92, 93, 97, 118, 119, 125, 161; in Kofyar homeland, 125–127; Tiv, 95, 188, 213
Malthus, T., 28
market and markets: appearing in ungwa, 176, 177; early Kofyar participation in, 80, 81; effect on crop choices, 105, 114; effect on settlement, 18–20, 41, 178; effects of, 10, 14, 17, 27, 34, 41, 80, 81, 86, 186, 212; efficiency, 202; geographical models of, 10, 14, 17, 176; incentives for intensification, 34, 40; participation by Kofyar and Tiv, 212; *soe loetuk* (eating the market), 176; travel to, 80, 81, 176–178; vs. subsistence economy, 14, 15, 19
mar muos (festive labor parties), 62, 63, 113, 115, 205, 209
Marx, K., 54, 186
mengwa (mai ungwa), 100: defined, 99
Merniang: bush farming, 81–88, 144; construed as a tribe, 68, 72, 83, 205; geographical area, 60, 80, 197; population density in, 60; *sargwat*, 67, 72, 99, 144–147, 151, 152, 205, 210; tax rate, 81, 83

Mesoamerica, settlement patterns, 6, 19, 21, 34, 51–53, 94
Middle East, settlement patterns, 16, 19
Mirriam. *See* Merniang
Mwahavul: on frontier, 152, 153; in homeland, 69, 71

Namu, 80, 179: character of, 3, 176; chief of, 86, 99, 100, 163; climate, 76; cultivated perimeter, 3, 80, 85, 86; cultural affiliation of, 77, 206; district population, 101; early Kofyar migration to, 3, 80, 85, 86, 88, 90; ethnic groups in, 3; history of, 59, 77, 206; Kofyar living in, 119; Muslim community of, 99, 119; raided, 59, 77, 206; travel to, 177, 178
Namu Sand Plains, 75, 76, 160, 165, 172
Netting, R. McC.: definition of Kofyar, 69; on households, 24, 62, 111, 205; Kofyar fieldwork, 8, 60, 80, 89, 90, 94, 125, 144, 146; on smallholder agriculture, 25, 34
Njak, 60, 77, 82–85, 205, 206
North America, settlement patterns, 6, 7, 21, 22, 42, 48, 51, 96, 140–143, 147, 186, 187, 201: Canada, 26, 96; Delaware Valley, 190–195; of freed slaves, 56; Great Plains, 23, 25, 34; Midwest, 41; New England, 23; Ozarks, 16; Southeast, 23, 24; Southwest, 24, 52

partial equilibrium analysis, 13, 18, 88
population density: correlated with log of agricultural intensity, 35; and cultivation radius, 46; and dispersal, 47–49; on frontier, 80, 101–103, 114; Kofyar homeland, 60; and production, 29, 30, 43
precipitation, 76
production concentration, 30, 33, 40: defined, 29
proximity-access principle, 22, 43: and agricultural fields, 43, 48, 49, 55, 121, 128, 182; defined, 14; effect on nonagricultural features, 18, 176; and interaction between settlements, 22, 121, 183

reciprocal labor. *See* agricultural collaboration; *wuk*
religion: effects on settlement, 119, 125–127, 210; *wumulak*, 65. *See also* Islam
roads, effect on settlement, 4, 20, 122, 162, 177–180, 209–212
"rules" of settlement, 9, 28, 181–186: social aspects, 7, 8, 184
R value, 104, 208

Sabon Gida, 93, 166, 213
sargwat (warfare alliance), 72, 139, 144–146, 152, 156, 157: defined, 67; endogamy, 69–72
satellite settlements, 49–51, 128
settlement evolution models: Bylund, 20, 24; Christaller, 19; Flannery, 6–8, 21, 120, 182; Hudson, 10, 11, 22–25, 120, 201; Morrill, 20; Siddle, 19
settlement fixation, 42: building costs, 44
settlement gravity, 120, 123, 143, 183: defined, 44
settlement spacing, 7, 123, 124, 182
sexual division of labor, 108, 110, 164
Shendam, 59, 68, 74, 76, 77, 80, 84, 206: chief of, 81, 86
shifting cultivation, 29, 36, 38–41,

45, 46: adopted by Kofyar, 83–90, 173; on American frontier, 41, 42, 190, 191; in Europe, 190
simulations, 7, 20, 168
social organization of labor. *See* agricultural collaboration; households; locational optimization, factors overriding; sexual division of labor
social physics models, 22, 120
sociosettlement units: on frontier, 88, 96, 114, 138, 183; in Kofyar homeland, 63–69, 73
soils: effects on settlement, 74, 83, 86, 88, 99, 164, 172, 173, 184, 185, 212; farmers' categories, 21, 184; sensitivity, 39, 165, 172
substantive rationality, 185, 186
Sura. *See* Mwahavul
symbolic aspects of settlement, 56

taxation, 81, 207: and ethnic subordination, 68, 72, 73, 151
territorial analysis, 15–17, 134, 138, 201
thresholds: Chisholm estimate, 17; 700 m, 92, 123, 132–139, 161, 164, 184, 210
Tiv, 84, 85, 92, 95, 97, 158, 207: Asumeku village, 213; Duwe, 85, 97, 187, 188, 207; language, 188; market participation, 212; settlement and sociopolitical organization, 187–189, 194; Tarkumburu village, 213. *See also* Kofyar, compared to Tiv
toenglu (neighborhoods), 3, 65, 151: on frontier, 151, 152; in homeland, 205
travel and transport costs, 21, 42, 49, 50, 203, 209, 210: effects on farm shape, 121–123; travel time, 14, 16, 41; trips to own farm, 121, 209. *See also* agricultural movement; locational optimization
tribe, 9, 64, 205, 206: colonial creations, 67, 68; Fried on, 68; "inclusive tribe," 69, 153; Kofyar, on frontier, 72, 144, 147; problems with term, 206. *See also* ethnicity
Turner, B., 29, 35, 202
Turner, F., 140, 141, 192

ungwa (settlement wards): defined, 3, 97; ethnicity within, 97, 145; evolution of boundaries, 96, 97, 146, 147, 151, 210; labor mobilization within, 4, 115, 117, 131–136, 209; political organization, 99, 100, 151
Ungwa Kofyar, 130, 134–138, 151, 177, 199, 208, 210, 212: agricultural intensification in, 108, 109, 115, 131, 133; settlement history of, 114, 115, 146, 169; *toenglu* in, 151, 152

von Thünen, H., 10, 13, 14, 42, 43, 121, 135, 138, 195: concentric land use in Africa, 46

warfare alliances. *See sargwat*
water: effects on settlement pattern, 16, 114, 122, 159–164, 184, 211; harvesting, 163; table, 160; transportation of, 211
work catchments, 123, 131–134, 137
wuk (reciprocal labor groups), 62, 151, 152, 208

About the Author

Glenn Davis Stone was educated at Northwestern University (B.A., 1977) and the University of Arizona (M.A., 1982; Ph.D., 1988). He is an anthropologist whose research on human ecology has embraced both archaeology and ethnography.

His interests initially centered on North American prehistory, and he conducted excavations on hunter-gatherer and agricultural sites in the eastern woodlands and the Southwest. He later developed more general theoretical interests in indigenous agriculture and settlement patterns. His articles have appeared in *American Anthropologist, American Ethnologist, Human Ecology, Journal of Anthropological Archaeology, Current Anthropology, World Archaeology, Journal of Anthropological Research,* and *Expedition,* and in edited books on archaeology, sociocultural anthropology, and geography.

He has twice conducted field research on agriculture and settlement in Nigeria, including an extended period in 1984–85 (in collaboration with R. McC. Netting and M. P. Stone), from which much of the information in this book is derived.

Stone has received multiple grants from the National Science Foundation and from the Wenner-Gren Foundation for Anthropological Research (including a Hunt Fellowship) and a Weatherhead Fellowship at the School of American Research in Santa Fe. From 1988 to 1995 he taught, as an assistant professor and associate professor, in the Anthropology Department at Columbia University in New York. Since 1995 he has been an associate professor in the Anthropology Department at Washington University in St. Louis, where he teaches courses on Africa, ecological anthropology, and computer/quantitative methods in social science.